10 0494744 3

D1458904

Neurotransmitters and Drugs

Neurotransmitters and Drugs

Third edition

Zygmunt L. Kruk

Reader in Neuropharmacology
Department of Pharmacology
Queen Mary and Westfield College
University of London

Christopher J. Pycock

Senior Registrar in General Medicine
Derriford Hospital, Plymouth

CHAPMAN & HALL
London · Glasgow · Weinheim · New York · Tokyo · Melbourne · Madras

Published by Chapman & Hall, 2-6 Boundary Row, London SE1 8HN, UK

Chapman & Hall, 2-6 Boundary Row, London SE1 8HN, UK

Blackie Academic & Professional, Wester Cleddens Road, Bishopbriggs, Glasgow G64 2NZ, UK

Chapman & Hall GmbH, Pappelallee 3, 69469 Weinheim, Germany

Chapman & Hall USA., 115 Fifth Avenue, New York, NY 10003, USA

Chapman & Hall Japan, ITP-Japan, Kyowa Building, 3F, 2-2-1 Hirakawacho, Chiyoda-ku, Tokyo 102, Japan

Chapman & Hall Australia, 102 Dodds Street, South Melbourne, Victoria 3205, Australia

Chapman & Hall India, R. Seshadri, 32 Second Main Road, CIT East, Madras 600 035, India

1004 947 443

First edition 1979
Second edition 1983
Reprinted 1987
Third edition
Reprinted 1993, 1995

© 1979, 1983, 1991 Zygmunt L. Kruk and Christopher J. Pycock

Typeset in 10.5/12pt Sabon by EJS Chemical Composition, Midsomer Norton, Bath, Avon
Printed in Great Britain by St Edmundsbury Press Ltd, Bury St Edmunds, Suffolk

ISBN 0 412 36100 0(HB)   0 412 36110 8(PB)

A Catalogue record for this book is available from the British Library

Library of Congress Cataloging-in-Publication Data available

∞ Printed on permanent acid-free text paper, manufactured in accordance with ANSI/NISO Z39.48-1992 and ANSI/NISO Z39.48-1984(Permanence of Paper).

Contents

Preface

This book is intended for students of medicine, pharmacy and other biological disciplines, who want to have a working knowledge of the mechanisms of action, uses and adverse effects of drugs which modify the activity of neurotransmitters in the peripheral and central nervous systems. It is suitable for undergraduates and for post-graduates on taught higher degree courses and diplomas.

New information and concepts have been incorporated into the text as appropriate, and references have been updated. Excitatory and inhibitory amino acids are considered in two chapters, and the last chapter of the first two editions (which considered drugs which do not interact selectively with neurotransmitters) has been subsumed into other sections.

The third edition follows the tried format of previous editions. Following a chapter which introduces the biology and pharmacology of neurotransmission, subsequent chapters deal with synthesis, storage, release, receptors and inactivation of individual neurotransmitters, together with a consideration of therapeutic uses and mechanisms of adverse effects.

Z.L. Kruk and C.J. Pycock
1991

1 Neurotransmission: sites at which drugs modify neurotransmission

The idea that nerves may communicate with other cells by releasing small quantities of chemicals at their junctions may have arisen from observations of the effects of poisons on animals. It was found that some poisons could mimic the effects of stimulating certain nerves, and it must have occurred to somebody that nerves release chemicals in response to stimulation. Histological studies showed that there is always a gap between the nerve ending and the target tissue, and that this gap must be crossed if the signal from the nerve is to reach its target.

Otto Loewi provided the first evidence for the actual release of a chemical in response to activation of a nerve. Using perfused frog hearts, he showed that a substance was released into the perfusion fluid when the vagus nerve to the heart was stimulated, and the heart slowed. If the perfusion fluid was passed into a second heart which was free of nervous stimulation, then this heart also was slowed. Loewi concluded that, when the vagus nerve was stimulated, a chemical which was released slowed the heart, and that this chemical passed into the perfusion fluid and acted to slow the second heart. More refined techniques were subsequently introduced to demonstrate this process in many organs and tissues. The process has been named **neurochemical transmission**, and the chemicals released have been called **neurotransmitters**.

Several chemicals that act as neurotransmitters have been identified, but not all substances that are found associated with nerves and are able to alter nervous activity are neurotransmitters.

The criteria by which it is established that a substance acts as a neurotransmitter are as follows.

1. The substance must be synthesized within the neurone from which it is released. Enzymes and substrates for synthesis must be present in the neurone.
2. The substance must be present in the neurone from which it is released. A storage mechanism exists for many neurotransmitters.
3. Calcium-dependent release appears to occur with all neurotransmitters. Such release must be shown to occur following physiological stimulation of the appropriate neuronal pathway.
4. A synthetic neurotransmitter applied exogenously must mimic the actions of the true transmitter when the latter is released in response to physiological or electrical stimulation. The exogenously applied substance must behave identically in every regard to the endogenous neurotransmitter in respect of the potentiation by inhibitors of enzymes of inactivation or re-uptake blockers, antagonism by competitive receptor blockers or physiological antagonists and electrical phenomena such as reversal potentials in the postsynaptic tissue.
5. There must be a mechanism for rapid termination of the action of a released neurotransmitter. The exogenously applied substance must be inactivated by the same mechanism as the true neurotransmitter.

1.1 THE NEURONE

The neurone or nerve cell is the fundamental cellular communication unit of the nervous system. It consists of the nerve cell body or perikaryon, which contains the nucleus, endoplasmic reticulum, Golgi apparatus and other components needed for synthesis of proteins and maintenance of intermediary metabolism. From the perikaryon there are one or more long outgrowths called axons, and more numerous shorter processes known as dendrites. The dendrites are usually considered to be concerned with receiving nervous impulses by means of specialized receptors, while axons are generally considered to be concerned with sending messages from the neurone to more distant structures located at the axon terminals. This generalized and much simplified structure of the neurone is shown in Fig. 1.1.

The process of transferring information from one end of a neurone to the other end is electrical, and detailed descriptions of these processes are to be found in texts of neurophysiology. Transfer of information from the neurone to a neuro-effector tissue is achieved by means of release of chemical(s) from the nerve terminal. The chemical diffuses across the synaptic cleft between the nerve terminal and the neuro-effector tissue and combines with the recognition site (the receptor). This process of chemical neurotransmission is susceptible to the action of drugs at several stages of the sequence, and this is the subject of the rest of this chapter.

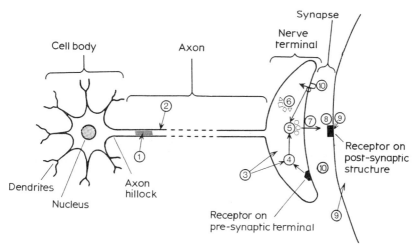

Figure 1.1 Sites at which drugs can modify neurotransmission. 1, axonal transport; 2, axonal membrane; 3, precursor availability; 4, synthesis; 5, storage; 6, intracellular organelles; 7, release; 8, receptors (pre- or post-synaptic); 9, post-receptor mechanisms; 10, inactivation: enzymatic or uptake.

1.1.1 Communication between cells

Cells send messages to each other by releasing messenger or information molecules. Neurotransmitters act at targets close to the nerve from which they are released; hormones may act at sites close to the cell from which they are released (the paracrine systems), or if they are released into the systemic circulation they may act at sites quite distant from the site of release (the endocrine systems).

Neurotransmitters and hormones are classes of information molecules known as **primary messengers**. The target cells for primary messengers have specialized mechanisms for recognizing the messenger, and for transferring the message from the outer surface of the cell, into the cell. The recognition site is called the receptor, and this is linked to a transducer mechanism which may amplify and distribute the message within the cell target cell. Molecules generated by transducer mechanisms are called **secondary messengers**.

(a) Primary messengers

Neurotransmitter primary messengers are classified in several ways: two commonly used schemes are

1. By chemical structure.
2. By their speed of action.

Table 1.1 Neurotransmitter systems: fast and slow

FAST	Amino acids	Excitatory	Glutamic acid
			Aspartic acid
		Inhibitory	Glycine
			Gamma aminobutyric acid
	Amines	Excitatory	Acetylcholine (nicotinic)
			5-HT (5-HT3)
SLOW	Amines		Acetylcholine, dopamine, noradrenaline, histamine, 5-hydroxytryptamine (5-HT)
	Peptides		Enkephalins, substance P, bradykinin

Both schemes are used in this book, and there is significant overlap between them. The chemical structure scheme groups neurotransmitters into three main categories, namely amines, amino acids and peptides. The speed of action scheme divides neurotransmitters into fast, slow or very slow groups. Amino acids and some amines are included among the fast transmitters, while other amines and peptides form the slow or very slow group. The major examples are shown in Table 1.1. Division of neurotransmitters into fast or slow is based upon the time it takes for transduction of the signal in the target cell, and this is dependent on the transduction mechanisms. **Fast transduction mechanisms** operate in the millisecond time domain, and they operate by direct opening of ion channels. **Slow transduction mechanisms** operate in the second/minute/ hour time domains, and work by activation of regulatory proteins and/or enzymes which generate secondary messengers.

1.2 AXONAL TRANSPORT

Axonal transport is a general term which refers to bulk axoplasmic flow, and to the specialized microtubule system found within axons. Materials are transferred along the axon by these processes to and from the nerve-cell body and nerve terminals. Axonal transport is needed since the nucleus which holds the genetic information for making enzymes is frequently far from the terminals at which the enzymes work. Axoplasmic transport both moves the enzymes necessary for transmitter synthesis and in the case of certain transmitters or neuromodulators, transports the active molecules themselves or their precursors. Some transmitter synthesis might occur during axonal transport, but this is not believed to be a major contribution to the total neurotransmitter found in the nerve terminal. The enzymes and organelles needed for the metabolic activity of the nerve are also carried by axoplasmic flow to the nerve terminal. Axonal transport is well illustrated by experiments in which axons are ligated (tied), and materials such as enzymes, granules and neurotransmitter accumulate on the nerve-cell body side (proximal side) of the ligature.

Substances which non-selectively interfere with axonal transport include vinblastine, vincristine and colchicine, all of which affect spindle formation in dividing cells and, as these structures have some features in common with neurotubules, they are also disrupted. Whereas these compounds have chemotherapeutic applications, they are only used to prevent axonal transport experimentally. Neurotoxicity due to neuro-tubule damage is a side effect of treatment with these compounds.

How molecules move along axons is not clearly established. The microtubule and axon fibre structure appears to be permanent, and movement of molecules may be achieved by mechanisms similar to those which underlie chromatography.

1.3 NEURONAL CELL MEMBRANES

The cell membrane of nerve cells (in common with most cells) is only semipermeable to ions. The uneven distribution of ions between the cytoplasm of a neurone and the extracellular fluid results in chemical concentration gradients across the cell membrane. As ions carry an electrical charge, an uneven distribution of ions on either side of a cell membrane results in an uneven charge distribution.

There is an uneven distribution of Na^+, Ca^{2+}, K^+, Cl^- and $proteins^-$, inside and outside cells. Intracellularly, there is a greater concentration of K^+ and $protein^-$; extracellularly there is a greater concentration of Na^+, Ca^{2+} and Cl^-. While K^+ can move relatively freely across the nerve cell membrane, the membrane is relatively impervious to Na^+. Diffusion of K^+ out of the cell leads to a net negative charge within most cells; in many nerve cells this is in the range -60 to -90 mV, and this is the resting potential of the cell.

Entry of positively charged ions into the cell (Na^+ and/or Ca^{2+} moving down their chemical concentration gradient) will result in depolarization: this is an excitatory effect. Egress of K^+ from the inside of the cell, or entry of Cl^- into the cell (both types of ion moving down their chemical concentration gradient) will result in hyperpolarization of the cell and is an inhibitory effect. Full descriptions of the ionic basis of membrane potentials can be found in physiology texts.

1.3.1 Ion channels

The nerve cell membrane is a double layer of lipid and protein, and located in this membrane, and spanning the double layer are specialized proteins called ion channels which (when open) link the cytoplasm of the cell with the extracellular fluid. Normally, ion channels are closed, but they can be opened in response to specific stimuli, namely either changes in cell membrane potential (potential or voltage regulated ion channels) or in response to the binding onto a specialized part of the ion channel (the receptor) of a specific chemical ligand (ligand or receptor operated ion channels; these are discussed further in section 1.9). The structure of ion channels is illustrated in Fig. 1.2.

Ion channels remain open for microseconds or less and they are capable of undergoing sequences of opening and closing tens of thousands of times each second. Four major groups of ion channel have been described, namely Na^+/K^+ channels, Ca^{2+} channels, Cl^- channels and K^+ channels. In specific locations, all appear to be capable of being regulated by either voltage sensors, ligands or intracellular messengers. In axonal membranes, voltage regulated ion channels are selective for small cations (i.e. Na^+ and K^+), thus opening of these channels leads to entry of sodium ions and depolarization of the cell membrane; egress of K^+ leads to repolarization. Propagation of the depolarization (action potential) occurs along the axon because voltage regulated ion channels in the adjacent part of the membrane become activated by the potential changes.

At axon terminals from which neurotransmitter is released, the depolarizing current is carried through voltage regulated Ca^{2+} channels. The change in ion is highly significant as calcium ions are essential for

---

Figure 1.2 Voltage regulated and receptor regulated ion channels consist of four or five protein sub-units, each consisting of between 500 and 800 amino acid residues. A single sub-unit is shown in (a), and 5 sub-units forming an ion channel are shown in (b). (a) A general scheme for a sub-unit is that there are six transmembrane spanning α-helices (numbered 1–6) joined by amino acid sequences in the extracellular space and within the cytoplasm. Helices numbered 1, 2, 3 and 5 form the outer part of the ion channel, those numbered 4 and 6 form the inner lining of the channel. (b) Electron microphotographs show that sub-units form a rosette when viewed from the extracellular surface, with the ion channel located in the centre. Binding sites for agonists and antagonists, and possibly the voltage sensor(s) are located on the extracellular surface. (c) The rosette depicted in (b), viewed in transverse section. Analysis shows that there is a ring of charged amino acids forming a 'gate' within the ion channel, which prevents movement of ions when the channel is closed. Occupation of the recognition site by an agonist (a molecule which activates the ion channel), or activation of the voltage sensor, causes rotation of helices 4 and 6 such that they move to the positions indicated by the arrows in (b). The ion channel is depicted in the open state by the dashed line (also indicated by the arrows).

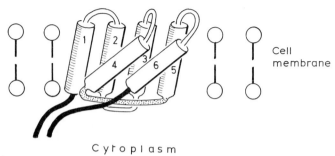

Extracellular space

Cell membrane

Cytoplasm

(a)

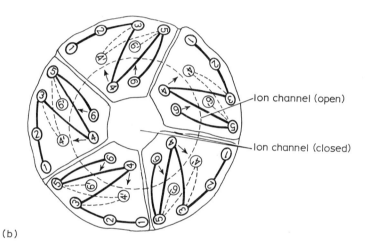

Ion channel (open)

Ion channel (closed)

(b)

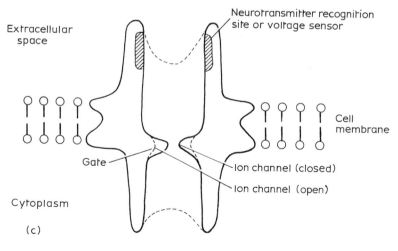

Neurotransmitter recognition site or voltage sensor

Extracellular space

Cell membrane

Gate

Ion channel (closed)

Ion channel (open)

Cytoplasm

(c)

release of neurotransmitter. Ca^{2+} entry into the axon terminal triggers release of transmitter, and this process links and transduces the electrical message conducted along the axon, to the chemical message carried by the released transmitter.

1.4 PRECURSORS

Two types of precursor appear to be used as sources of neurotransmitter. The transmitter may be synthesized in the nerve terminal from which it is released. At such nerve terminals there is an active transport system in the cell membrane which carries the precursor into the nerve from the extracellular space. Examples of such precursor-uptake systems include **tyrosine** uptake into noradrenergic neurones, **tryptophan** uptake into 5-hydroxytryptamine neurones and **choline** uptake into cholinergic neurones. Increased precursor availability is the basis of some forms of therapy. Inhibitors of the precursor uptake systems are of experimental interest.

In those neurones in which the transmitter is not synthesized in the nerve terminal, a larger precursor molecule may be synthesized in the nerve-cell body and then carried by axonal transport to the nerve terminal. The large precursor molecule is broken down enzymatically into a smaller molecule, which is then released. This appears to be the mechanism by which peptides are brought to nerve terminals. Inhibitors of axonal transport are the only drugs which can modify the availability of these precursors (section 9.2).

1.5 SYNTHESIS

Enzymes found in nerve terminals, together with any cofactors and necessary ions, catalyse the synthesis of neurotransmitter from precursor. Depending on the characteristics of the enzymes involved, it may be possible to speed synthesis by increasing the availability of substrate; intraneuronal mechanisms which control the rate of synthesis may limit the effectiveness of such procedures. Inhibitors of the enzymes of synthesis will decrease the amount of transmitter available for release. Such inhibitors are used mainly for research purposes.

1.5.1 Control of synthesis

The rate of synthesis of many transmitters is closely linked to the rate at which they are released; the rate of arrival of nerve impulses at the nerve terminal can speed or slow synthesis to keep pace with release.

(a) End-product inhibition

In many instances, end-product inhibition of the rate-limiting enzyme is of importance. High concentrations of **noradrenaline** in noradrenergic neurones, for example, inhibit the enzyme tyrosine hydroxylase, which determines the rate of noradrenaline synthesis.

(b) Presynaptic receptors

Presynaptic receptors are believed to detect the concentration of neurotransmitter present in the synaptic cleft and to control the rate of synthesis and release of transmitter appropriately. This is one of the mechanisms which serve a homeostatic role in maintaining nervous activity within set limits.

Presynaptic receptors may recognize either the neurotransmitter released by that neurone or other neurones. In the former case the receptors are called *autoreceptors*; in the latter case they are called *heteroreceptors*.

(c) Availability of precursor

If the rate-limiting enzyme is not normally saturated by substrate, then increasing the substrate will result in greater synthesis of neurotransmitter.

1.5.2 Turnover of neurotransmitter

This is a term used to describe biochemical measurements (usually of neurotransmitter metabolites or precursors) made in an attempt to get an

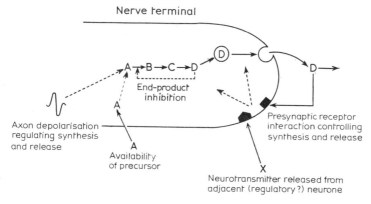

Figure 1.3 Mechanisms regulating synthesis and release of neurotransmitters.

'estimate' or 'index' of the activity of a neurone in relation to its utilization of neurotransmitter. The best measure of utilization of functional neurotransmitter is to measure release of transmitter. This continues to be a technically difficult procedure; electrochemical detection methods at microelectrodes allow real-time measurement of release of some monoamine transmitters. Caution should be exercised in extrapolating from biochemical measurements of neurotransmitter metabolites to poorly defined notions of 'functional state' of a neurotransmitter system.

1.6 STORAGE

There appear to be several storage forms of neurotransmitters, and a particular transmitter may be stored in more than one form. Evidence for multiple forms of storage of individual transmitters comes from anatomical, biochemical and pharmacological experiments; it is frequently not possible to obtain agreement from different experimental approaches as to which forms of transmitter storage serve which functions. What does seem to be agreed is that newly synthesized transmitter is generally released in preference to that which has been stored. Certain neurotransmitters appear to be stored in vesicles within nerve terminals. Vesicular storage has not been demonstrated for all transmitters, and indeed certain nerve terminals do not appear to contain vesicles. In such neurones, the transmitter is presumably stored in a different form – for example, in solution in the cytoplasm.

If neuronal tissue is homogenized by mechanical disruption in an isotonic medium, and the homogenate is centrifuged under appropriate conditions, it is possible to obtain fractions of the homogenate which contain high concentrations of neurotransmitter. Examination of the neurotransmitter-rich fractions under the electron microscope shows that they are composed of what appear to be broken-off nerve endings, some of which contain synaptic vesicles. Such broken-off nerve endings are called **synaptosomes**, and it is believed that they represent presynaptic nerve terminals. Significant amounts of neurotransmitter have been detected in other fractions of homogenates, and this has been taken to indicate that transmitter may be stored in other subcellular structures.

Some neurotransmitters appear to be stored in complexes which include synthesizing enzymes, structural proteins and metal ions. Drugs which decrease the stability of the storage complex may result in disruption of storage complexes, and allow transmitter to diffuse into the cytoplasm. A major function of specialized storage complexes of neurotransmitter is believed to be the protection of the transmitter from destruction by enzymes within the nerve terminal; if storage is disrupted, then transmitter will be destroyed. This may inhibit the neurotransmitter function of the nerve as a result of the decreased availability of transmitter for release.

1.7 ORGANELLES AND ENZYMES

Organelles and enzymes in the nerve ending maintain processes necessary for both the metabolic and the neurotransmitter activity of the cell. The absence of energy substrates – for example, following exposure to metabolic poisons – will lead to decreased transmitter function. The inhibition of enzymes concerned with transmitter synthesis or destruction will have obvious consequences. As pointed out above, neurones maintain a controlled level of transmitter synthesis. Transmitter which is in excess of that needed for release, and which cannot be stored (because storage capacity is exceeded), is usually inactivated intraneuronally by enzymes, resulting in a product which is biologically less active. Inhibition of such enzymes may increase the amount of transmitter stored in the nerve ending. Mitochondria are the structures in which are found the electron transport systems responsible for oxidative phosphorylation and synthesis of high energy compounds such as ATP. Monoamine oxidase (MAO) is located in mitochondria.

The endoplasmic reticulum is a system of intracellular membranes. Located in the membranes are active transport pumps, ion channels and probably receptors. Among the important functions of the endoplasmic reticulum is the regulation of Ca^{2+}; many proteins are critically sensitive to the concentration of Ca^{2+} in the cytoplasm, and the concentration of 'free calcium' in the cytoplasm is strictly regulated. Uptake into the endoplasmic reticulum, active and exchange transport out of the cell, binding to calcium binding proteins, and release from endoplasmic reticulum by the secondary messenger IP3 (section 1.9.2) are all crucial to calcium homeostasis.

1.7.1 Kinases, phosphorylation and phosphatase

An important mechanism for activation of enzymes is phosphorylation, and this is achieved by means of enzymes called **kinases**. Secondary messengers (section 1.9.2) (cyclic AMP, IP3, DAG, Ca^{2+}) lead to activation of kinases by phosphorylation. The phosphorylated kinases alter the activity of enzymes and structural proteins, leading to cellular response. Examples of these mechanisms include protein kinase C, which is a cell membrane bound enzyme activated by Ca^{2+} and by DAG; when activated protein kinase C (or more accurately protein kinase Cs, for there are many iso-enzymes) lead to phosphorylation of nicotinic ACh receptors, and voltage regulated sodium and calcium ion channels. The functional significance of these phosphorylations is not clear; it has been suggested that phosphorylation of ACh nicotinic receptors leads to internalization of nicotinic receptors in the membrane (which could lead to desensitization). Phosphorylation of voltage regulated Ca^{2+} channels may be involved in the establishment of LTP (long term potentiation), a process associated with learning and memory (section 8.4.3). c-AMP regulated phosphorylation of phosphorylase kinase leads to glycogenolysis in liver, and phosphorylation of tyrosine hydroxylase (the rate limiting step enzyme in the synthesis of catecholamines) determines in part the activity of this enzyme. When catecholamine axon terminals are electrically stimulated then tyrosine hydroxylase becomes phosphorylated by a kinase regulated by neuronal activity. Phosphatases are enzymes which dephosphorylate enzymes and other proteins; dephosphorylation of tyrosine hydroxylase leads to decreased activity of that enzyme.

1.8 RELEASE

At least two processes may operate during the release of neuro-transmitter. Some transmitters appear to be released by a process of exocytosis which involves the fusion of vesicular membrane with the presynaptic nerve membrane. Other transmitters are released by less well-defined processes; diffusion through presynaptic membranes and passage through special channels have been suggested. The released transmitter is then free to diffuse across the synaptic cleft.

All neurotransmitter release (and secretory mechanisms in general) are dependent on changes of intracellular Ca^{2+}, which may enter either through voltage regulated ion channels (section 1.3.1) or be released from intracellular stores (section 1.7).

Release of transmitter from vesicles is achieved by mechanisms which are poorly described, but it has been suggested that movement of cytoskeletal and vesicular proteins may facilitate the apposition and fusion of vesicular and cell membranes (thus leading to exocytosis). Some support for such a scheme comes from observations of release of transmitter in relation to the activity of protein kinase 2. Protein kinase 2 is an enzyme which phosphorylates cytoskeletal proteins; its activity is controlled by a calcium binding protein (calmodulin). When calcium enters the cell, it is bound to calmodulin, which in turn binds to and activates protein kinase 2. If the activity of protein kinase 2 determines the activity of cytoskeletal and vesicular proteins, and hence release of transmitter, then a link between calcium entry and release may have been established. Such an idea is attractive as it could account for rapid release of transmitter (when Ca^{2+} entry is rapid through voltage regulated ion channels), and slower modulation of release (by autoreceptors), when changes in calcium concentration in the cytoplasm may be achieved by slower (secondary messenger dependent) processes.

1.8.1 Drug induced modification of transmitter release

1. Disruption of storage may cause liberation of transmitter inter-neuronally, where it may be destroyed by enzymes; no receptor activation will occur following such disruption of stores.
2. Drugs are able to displace neurotransmitter from its stores and bring about release into the synaptic cleft. Such drugs are known as indirect receptor agonists and their action is not calcium dependent.
3. Drugs can be taken up into the nerve terminal and be substituted by exchange for the true transmitter molecules; such molecules may have no biological activity at receptors. Such compounds have been referred to as false transmitters (section 3.1.3).
4. Drugs with local anaesthetic activity which are selectively taken up by some nerves may prevent release of transmitter, by stabilizing the nerve membrane through which release usually occurs; they may do so by preventing the process of calcium entry or secretion.
5. Auto- and heteroreceptor activation can lead to either increased or decreased release of neurotransmitter.

Modulation of neurotransmitter release by autoreceptors and by changes in axon traffic may be regulated by phosphorylation of cytoskeletal proteins which regulate the location and movement of synaptic vesicles. Protein kinase 2 is an enzyme whose activity is regulated by the concentration of calcium in the nerve terminal. Increases of calcium concentration in the nerve terminal caused by either its entry through voltage regulated channels, or as a result of secondary messenger dependent processes, leads to activation of calmodulin, a calcium binding protein which (when it has calcium bound to it) activates many protein kinases.

1.9 RECEPTORS

Receptors are proteins found in membranes which can selectively bind messenger molecules and transduce the chemical signal into a response in the target cell. The majority of receptors are cell surface receptors. Intracellular receptors are found on the membrane of the nucleus (receptors for steroid hormones) and there is evidence that there are specific recognition sites for intracellular messenger molecules on the surface of the endoplasmic reticulum and in the cytoplasmic regions of some ion channels. Our knowledge of intracellular receptors is limited and discussion of receptors is restricted here to cell surface receptors, which can be divided into two large super families. Receptor regulated ion channels (Fig. 1.2) are used by mechanisms where good time

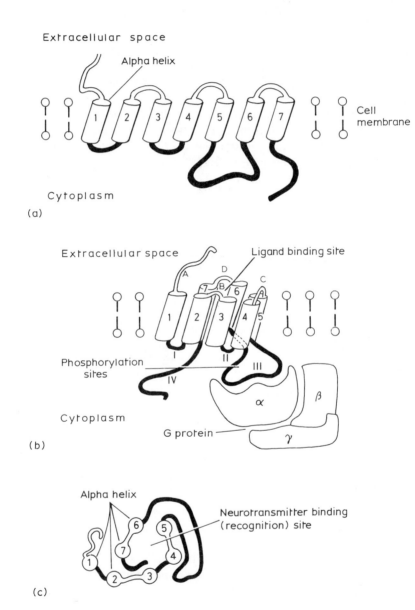

discrimination is essential and they are able to follow high frequency events with great accuracy (high fidelity), and are used by mechanisms where fast and accurate responses are essential.

The second family of receptors use transduction mechanisms involving regulatory or 'G' proteins (Fig. 1.4). The G proteins control enzymatic or other processes in the cell membrane, and the response time of these receptors is slower, typically taking seconds or minutes to become fully effective.

1.9.1 G proteins/regulatory proteins

G proteins consist of three sub-units; alpha, beta and gamma. They act as a link between the ligand binding site on the cell surface and the effector mechanism or enzyme embedded on the cytoplasmic side of the cell membrane.

When a binding site is occupied by an agonist, the affinity of the alpha sub-unit for GTP increases. When GTP is bound to the alpha sub-unit the affinity of the alpha sub-unit for its links with the beta gamma sub-unit is reduced. Under these circumstances the beta gamma sub-unit dissociates

---

Figure 1.4 The structure of G protein regulated receptors has been deduced from amino acid sequence analysis which itself has been deduced from analysis of DNA base sequences. Our ability to determine secondary, tertiary and quaternary structure of proteins from amino acid sequences is still rudimentary. Some sequences of the amino acids are associated with particular types of protein structure: we can recognize the amino acid sequences which give rise to α-helices and those which give rise to β-pleated sheets. This can be backed up by estimates of the hydrophobicity of amino acid sequences and taken together an educated guess is made at the location and possible function of sections of amino acid sequences of receptor peptides.

G protein regulated receptors consist of between 400 and 600 amino acids in the receptor protein. The model of the location of this receptor (parts (a) and (b)) in the membrane suggests that there are seven transmembrane regions (labelled 1–7) with the COOH group in the extracellular space and NH_2 ending in the cytoplasm. There are four extracellular domains (A, B, C, D) and four cytoplasmic domains (I, II, III, IV). The location of binding sites for agonists and blockers has been established for some neurotransmitters and this is illustrated. G proteins bind to the third intracellular amino acid loop (III), and phosphorylation of amino acids on this loop regulates sensitivity of some receptors as discussed in the text (section 3.4). The arrangement of the seven transmembrane spanning regions as viewed from the extracellular space is shown schematically (c).

Neurotransmitter molecules are believed to attach to specific amino acid residues (which form the neurotransmitter recognition site), which line a 'pit' made by the transmembrane spanning regions of the receptor molecule.

from the alpha sub-unit, and its affinity for its binding site on the effector enzyme is now increased. When the beta gamma sub-unit binds to the effector enzyme it modifies the activity of the enzyme (Figure 1.5).

G proteins come in three different types. G_s stimulate effector enzymes; G_i inhibit effector enzymes and G_o have functions other than stimulation or inhibition which we do not understand. The best characterized enzymes which are regulated by G proteins are adenylyl cyclase and phospholipase C.

Activation or inhibition of enzymes by the beta gamma sub-unit is terminated by hydrolysis of the GTP bound to the alpha sub-unit. When the GTP is broken down to GDP then the affinity of the alpha sub-unit for the beta gamma sub-unit increases to the extent that the alpha beta gamma sub-unit now re-associates and the beta gamma sub-unit is removed from the enzyme. The cycle can now start again. In some systems the alpha subunit/GTP complex dissociates and activates the effector.

1.9.2 Secondary messenger systems

The two best described effector enzymes regulated by G proteins which catalyse the production of secondary messengers are adenylyl cyclase which generate cyclic 3'5-adenosine monophosphate (cyclic AMP, c-AMP), and phospholipase C, which generates inositol 1,4,5-trisphosphate (IP3), and diacyl glycerol (DAG). c-AMP and IP3 are water soluble, so they diffuse readily in the cytoplasm to their sites of action: c-AMP activates kinases (section 1.7.1); IP3 binds to intracellular receptors on the endoplasmic reticulum. The IP3 receptors mediate the opening of calcium ion channels in the endoplasmic reticulum and lead to an increase in the concentration of calcium in the cytoplasm, thus calcium itself can be regarded as a secondary messenger. Liberation of calcium into the cytoplasm activates many enzymes including some protein kinases and calcium binding proteins such as calmodulin. In addition, calcium can act on the cytoplasmic surface of some ion channels to alter their state of opening. Regulation of potassium ion channels by changes in intracellular calcium concentrations is important for regulating the degree of excitability of some cells.

Diacyl glycerol (DAG) (which is lipid soluble) can activate protein kinase Cs, enzymes which phosphorylate a wide range of structural and functional proteins.

In addition to regulating the activity of secondary messenger generating enzymes, G proteins can directly regulate the opening state of some ion channels. Beta gamma sub-units and in some experiments also GTP bound alpha sub-units have been found to directly regulate the opening of some potassium channels. Opening of K^+ channels leads to hyperpolarization of a cell (due to loss of positively charged ions from the

Table 1.2 Transmitters: receptors: transducers

Neurotransmitter	Receptor name	Transduction mechanism
Noradrenaline	Alpha 1 adrenoreceptor	Activates PLC
	Alpha 2 adrenoreceptor	Inhibits AC Gpr regulated K^+ channel
	Beta adrenoreceptor	Activates AC
Acetylcholine	Muscarinic receptor	Activates PLC inhibits AC Gpr regulated K^+ channel
	Nicotinic receptor	Intrinsic Na^+/K^+ channel
Dopamine	D1 receptor	Activates AC
	D2 receptor	Inhibits AC Gpr regulated K^+ channel
5-HT	5-HT 1 receptor	Inhibits AC
	5-HT 2 receptor	Activates PLC
	5-HT 3 receptor	Intrinsic $Na^+/K^+/Ca^{2+}$ channel
Histamine	H1 receptor	Activates PLC
	H2 receptor	Activates AC
	H3 receptor	?
GABA	$GABA_A$ receptor	Intrinsic Cl^- channel
	$GABA_B$ receptor	Gpr regulated K^+ channel Inhibits AC
Glycine	Glycine receptor	Intrinsic Cl^- channel
Glutamic acid	NMDA receptor	Intrinsic $Na^+/K^+/Ca^{2+}$ channel

Abbreviations: AC, adenylyl cyclase; PLC, phospholipase C; Gpr, G protein.

inside of the cell) and this is a means of altering cell excitability. Table 1.2 shows a selected list of receptors for some neurotransmitters and the transducer and associated effector mechanisms.

Presynaptic autoreceptors are found on nerve cell bodies and dendrites (somatodendritic autoreceptors) and on axon terminals. At somatodendritic sites, autoreceptor activation leads to decreased spontaneous or transynaptically regulated electrical activity. Stimulation of axon terminal autoreceptors generally leads to a decrease in the release of the neurotransmitter from that nerve terminal. At some sites, however, axon terminal autoreceptors facilitate transmitter release and at some somatodendritic sites autoreceptors also regulate transmitter release.

1.9.3 Definitions and terminology

Agonists are substances which initiate a response in the neuroeffector tissue. Tissues generally have a maximal response which cannot be

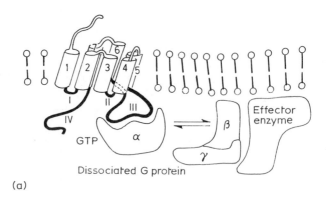

(a)

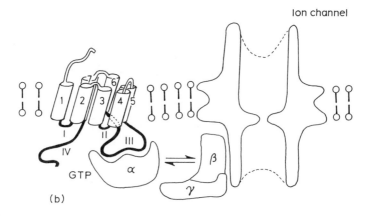

Ion channel

(b)

Figure 1.5 When an agonist molecule is attached to the ligand recognition site, affinity of the α-sub-unit of the G protein for GTP increases, and when GTP is bound to the α-sub-unit, the β- and γ-sub-units dissociate and attach to the effector enzyme (adenylyl cyclase or phospholipase C). The effector enzymes are adenylyl cyclase, which when activated catalyses the conversion of ATP to cAMP, and phospholipase C, which catalyses the lysis of plasma triglyceride to diacyl glycerol (DAG) and inositol phosphate, which undergoes further phosphorylation to 1,4,5 inositol trisphosphate (IP3) (a).

Some receptors (including some muscarinic M2 sites, some sub-types of the dopamine D2 receptor and some types of 5-HT 1 and GABA$_B$ sites) consist of a hybrid of G protein and ion channel regulated complex. The exact mechanisms of regulation of these sites is not clear, but it is suggested that G proteins adjacent to the ion channel are able to regulate the opening state of the ion channel by an intramembrane interaction. In the illustration (b), a β-γ-sub-unit dissociates and interacts with a recognition site on a K ion channel.

exceeded; substances which can initiate the maximal response are known as full agonists. A substance which initiates a response in the tissue but which cannot initiate the maximal response is known as a partial agonist. An alternative term for agonist is stimulant.

Substances which prevent an agonist from initiating a response are known as antagonists or blockers. Some antagonists have a limited capacity to initiate a biological response. This property of antagonists is referred to as partial agonist activity. In this book, the term receptor blocker will be used for antagonists acting at receptors for neurotransmitters.

Competitive antagonism is a term reserved for the process by which a receptor blocker prevents access of the agonist (neurotransmitter or drug) to the receptor but, if the concentration of agonist is increased, the competitive antagonist will be displaced and response will occur.

Physiological antagonism is a term reserved for the action of substances which act at different receptors, and bring about opposite effects in the same neuroeffector tissue. For example **noradrenaline** is a physiological antagonist for **acetylcholine**.

The potency of a drug refers to the dose or concentration of a drug needed to elicit a standard response; the more potent a compound, the less is needed to obtain the standard response. Potency can be used in relationship to any action of a drug. It gives no indication of the qualitative properties of a drug.

The affinity of a drug generally refers to the ease with which a drug becomes linked with its target site. It is a measure of the attraction of a drug for its target site, for example, a receptor or enzyme.

The efficacy or intrinsic activity of a drug is a measure of the ability of the drug to initiate a response. Its use is restricted to substances with agonist actions. Agonists possess both affinity and efficacy; receptor blockers possess affinity but lack efficacy; partial agonists possess affinity but a lesser efficacy than full agonists. In the case of agonists, potency is a function of both affinity and efficacy; in the case of receptor blockers, their potency depends only on affinity.

1.9.4 Changes in receptor sensitivity

Under well-defined conditions, changes in receptor sensitivity may occur, either as changes of affinity or as changes in receptor numbers. Following denervation, receptor supersensitivity occurs, and thus the response to exogenously applied agonists is greater than the pre-denervation maximum. Supersensitivity can also occur following prolonged periods of receptor blockade with drugs. It is believed that the adaptive process is one which allows function to be maintained despite either decrease in transmitter or decrease in number of available receptors. Receptor desensitization or subsensitivity can occur following prolonged stimulation of receptors by agonists. Again, this would seem to be an adaptive process in response to excess stimulation of receptors. Loss of total receptor number or decrease in affinity may be mechanisms of receptor subsensitivity, and this may occur in some forms of tolerance to drugs. When tolerance occurs acutely, following only a few applications of agonist, then it is called tachyphylaxis. Changes in receptor sensitivity may operate in tachyphylaxis but, with indirectly-acting agonists, it is possible that the exhaustion of transmitter available for release is an important mechanism.

1.9.5 Receptor (ligand) binding studies (Fig. 1.6)

Our ability to characterize drug binding sites has been greatly facilitated by the use of radioactively labelled drugs, which bind to cell surface receptors. The advantage of this method over pharmacological experiments in which tissue responses are measured in response to drug, is that ligand binding studies are quantitative, simple and quick. A drawback of these studies is that they do not measure a biological response, so to be useful, ligand binding studies must always be compared with quantitative and qualitative pharmacological experiments in which responses to the same set of drugs is assessed in experiments where a

---

Figure 1.6 Ligand binding studies. In (a), the radioactive ligand (closed circles) is shown occupying all possible binding sites. Specific sites (receptors) are depicted as 'U', non-specific binding sites as closed triangles. This represents total binding (12 sites occupied). In (b), the receptor sites are occupied by a non-radioactive competitive ligand 'X', so the radioactive ligand can only occupy the non-specific sites. This represents non-specific binding (8 sites occupied). By subtracting non-specific binding from total binding, the specific binding is measured (four sites).

The graphs (c), (d) and (e) show results obtained when varying concentrations of radioactive ligand are incubated with a fixed amount of protein. Specific binding is obtained from (c) by subtraction, and this is shown in (d). B_{max} is a measure when all specific sites are occupied; this is expressed as concentration of ligand bound/unit of protein. At half B_{max}, half of the binding sites are occupied, and this is a measure of the affinity of the ligand for the binding site. This concentration of ligand is called the affinity constant and is represented as K. In (e), the results are shown in a Scatchard plot, where the ratio of ligand bound: free is plotted against ligand bound. The slope of the line is now numerically equal to the affinity constant, and B_{max} is where the slope crosses the X-axis.

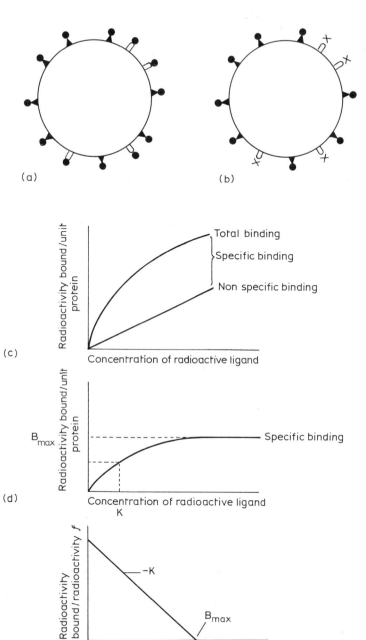

(a)

(b)

(c)

Radioactivity bound/unit protein

Total binding

Specific binding

Non specific binding

Concentration of radioactive ligand

(d)

B_{max}

Radioactivity bound/unit protein

Specific binding

Concentration of radioactive ligand

K

(e)

Radioactivity bound/radioactivity f

−K

B_{max}

Radioactivity bound/unit protein

biological response is measured. Ligand binding studies are carried out in a test tube which contains appropriate buffer, some purified cell membranes, and radioactive drug (ligand). The mixture is incubated for a standard time, and the membranes are separated from the incubation medium by either filtration or centrifugation. The radioactivity of the membranes is then measured by scintillation counting. Experiments show that the more radioactive drug that is added, the more radioactivity is found associated with the separated membranes. The radioactive drug is bound to two classes of site, namely the specific binding sites (receptors), and non-specific binding sites found on the cell membrane preparation used. An act of faith in these studies (which after 20 years of experience with the methods shows that it is largely justified), is that the specific (receptor) binding sites have a higher affinity for the drug than do the non-specific sites. This allows us to differentiate between the sites by careful use of drugs which are known from biological experiments to have a high affinity for the drug receptor. Incubating the radioactive drug with cell membranes as described above but in the presence of a non-radioactive drug which is a known competitive ligand at the receptor, leads to the specific (receptor) binding sites being occupied by the non-radioactive drug, while only the non-specific sites bind the radioactive drug (Fig. 1.6c–e). This leads to a measure of the amount of binding to the non-specific sites. If this non-specific binding is now subtracted from the total, what is left is a measure of the amount of radioactive drug bound to the specific or receptor sites. In addition to work in test tubes on isolated cell membrane preparations, ligand-binding studies can be performed in slices of tissue prepared for autoradiography; this allows direct location of receptors in tissues and assigning them to specific cell types.

1.10 INACTIVATION

Following release, a neurotransmitter would continue to stimulate receptors if mechanisms for terminating its action did not exist. Two main mechanisms have been identified.

Enzymatic inactivation of transmitter must occur in the synaptic cleft if other mechanisms are not to be involved. Examples of membrane-bound and soluble enzymes are known, and they are located in the synaptic cleft. Acetylcholinesterase is found on synaptic membranes.

Neuronal re-uptake is another mechanism for terminating the biological activity of neurotransmitter released into the synaptic cleft. The uptake process frequently occurs back into the neurone from which the transmitter was released; this is an active-transport system, which has a high affinity for the neurotransmitter. An advantage of re-uptake mechanisms into neurones is that the transmitter is available for release on a subsequent occasion. Inhibition of inactivation processes prolongs the availability of transmitter at the receptor, and such drugs have major therapeutic applications.

1.11 NEUROTRANSMITTERS OR NEUROMODULATORS

The criteria by which a substance is judged to be a neurotransmitter have been outlined above. Neuromodulators are substances which act on or within neurones to alter the responsiveness of the neurotransmitter function of the neurone. Thus, drugs which act at any of the sites reviewed above can be thought of as modulators of neurotransmitter action. In this text (and in the conventional wisdom of the neurosciences), the term neuromodulator is reserved for substances normally present in the body which modify the responsiveness of a neurotransmitter system by altering transmitter synthesis or release, or possibly by altering the sensitivity of postsynaptic receptors, or the activity of secondary messengers.

Recent work has led to the suggestion that many neurotransmitters can modify the activity of the neurones from which they are released, as well as the activity of other neurones. Distinguishing a neurotransmitter from a neuromodulator role is becoming more difficult and becomes a matter of semantics. The existence of more than one neurotransmitter in the same neurone (co-transmission) is discussed in section 9.3, and this is one example of neuromodulation. Others include the presynaptic receptors on noradrenergic nerve terminals (section 3.3.1).

1.12 NEUROTRANSMITTERS IN THE NERVOUS SYSTEM

The mammalian nervous system can be divided into two parts. The central nervous system (CNS) which lies within the skull and vertebral column, and which consists of the brain and spinal cord, and the peripheral nervous system which lies outside the CNS. The peripheral nervous system thus comprises the twelve pairs of cranial nerves which leave the brain directly, the 31 pairs of spinal nerves which leave at each segment of the spinal cord and intrinsic nervous systems as in the gut. The peripheral nervous system is in part concerned with transmitting information from peripheral sense organs and terminals to the central nervous system, and conveying instructions from the central nervous system to effector tissues.

There are two large subdivisions of the peripheral nervous system, namely the somatic nervous system, and the autonomic nervous system.

The somatic nervous system consists of the voluntary motor nerves to skeletal muscle, and sensory nerves. The motor nerves of the somatic system are under voluntary control. The autonomic nervous system (ANS) innervates the muscles, glands and blood vessels of the internal body organs. Autonomic nerves also serve a sensory function, conveying information from the body organs to the central nervous system. The autonomic nervous system is concerned with control of bodily functions which are not under direct voluntary control. It is thus a self-regulating system in the main, and the animal is not generally aware of many of the events and systems controlled by the autonomic nervous system.

The autonomic nervous system can be divided functionally and anatomically into two subdivisions, the sympathetic and the parasympathetic. The sympathetic nervous system is most active during stress, and it is responsible for co-ordinating responses of the internal organs and blood vessels to situations which involve fright, flight or fight responses. The major neurotransmitter in the sympathetic nervous system is **noradrenaline**, and its distributions and functions are described in Chapter 3.

The parasympathetic nervous system is most active during conditions of bodily rest, and it co-ordinates responses of the internal organs during rest, digestion of food and sleep. The major transmitter in the parasympathetic nervous system is **acetylcholine**, and its distribution and functions are described in Chapter 2.

It is important to be aware that the switching between dominance of the two divisions of the autonomic nervous system is not an 'all-or-none' mechanism. The two systems are always active, but under the appropriate conditions as mentioned above, either one or other of the subdivisions is dominant, and under 'normal' circumstances, there is a balance of sympathetic and parasympathetic nervous system activity which co-ordinates the activity of the internal body organs and blood vessels.

The discovery that the terminal neurones in the autonomic nervous system to internal organs release either acetylcholine or noradrenaline led to the idea that there are chemical pathways in the nervous system in general, not just in the autonomic system, but also in the other parts of the peripheral nervous system, and in the central nervous system. This has indeed proved to be the case, and with the development of more refined and more sensitive techniques for mapping chemical pathways, it has been possible to describe the points of origin and termination of neuronal systems that use the chemicals which are the subjects of Chapters 2–9. Such studies have allowed a better understanding of the nervous mechanisms which control specific functions, and have gone a long way towards explaining the mechanisms of action of many drugs.

1.13 NEUROTRANSMITTERS AND DRUGS

A multidisciplinary approach has led to the present day descriptions and understanding of the mechanisms of neurotransmission. Pharmacology is the study of the actions of chemicals (drugs) in animals. In the past 100 years this has changed, from a description of gross effects of chemicals in whole animals or isolated organs or tissues, to a description of the interactions of drug molecules at identified molecular sites in cells.

The use of drugs has always been important in studies of neurotransmission, and what were once the chemical tools for unravelling the mechanisms involved, have now become the therapeutic agents used to treat disorders of those biological systems. In the present context the mechanisms of neurotransmission are the biological system, and drugs are the chemical tools used to study and control the system.

1.13.1 How drugs can be classified

Classifications are used by specialists to enable them to place information in an ordered and readily retrievable system. No method of classification of any area of human knowledge is perfect, and thus multiple classifications of the same area of knowledge arise. Information about drugs is no exception and what is presented below is an attempt to indicate the ways in which drugs have been grouped together, and the advantages and limitations of individual methods. This is not an exhaustive exercise, but the main conventional categories are covered.

(a) Classification by chemical structure

This presumes that the user is familiar with the pharmacological properties of a chemical group. As these vary within a chemical group, details of exceptions to rules must be known.

(b) Classification by physiological or anatomical target

Drugs acting on a particular physiological system or anatomical organ will have many and diverse effects. A knowledge of the physiology of the system may allow predictions to be made about the effect of specific agents.

(c) Classification by stimulant or depressant actions

The target has to be defined, and the response must also be defined. It is frequently necessary to qualify this form of classification since exceptions abound, especially when different dose levels are considered.

(d) Classification by therapeutic use

If the pathology of the condition is well understood, then it is possible to predict what sort of agents are liable to be of benefit by a consideration of their mechanisms of action.

(e) Classification as psychotropic or non-psychotropic drugs

If the main intention of giving a drug is to bring about a change in mood or behaviour, then that compound is being used as a psychotropic agent. Effects on mood and behaviour are produced by many compounds which are not intended to do so.

(f) Classification by neurochemical interactions

Drugs can show a degree of selectivity for specific neurochemical systems, and thus one can refer to compounds as interacting with particular neurotransmitters. It may be possible, further, to say that specific processes unique to that neurotransmitter are being affected. If the functional role of that transmitter is known, it is possible to predict what the effect of a drug with a given action will be. It is this classification of drugs which has been utilized in this book.

FURTHER READING

Bowmer, C.J. and Yate, M.S. (1989) Therapeutic potential for new selective adenosine receptor ligands and metabolism inhibitors. *TIPS*, **10**, No. 9.

Boyer, J.L., Hepler, J.R. and Harden, T.K. (1989) Hormone and growth factor receptor-mediated regulation of phospholipase C activity. *TIPS*, **10**, No. 9.

British National Formulary. (BNF) New Edition about every 9 months. Short reviews of currently available medical treatments discussed in relation to physiological systems plus comprehensive listing of available medicines.

Catterall, W.A. (1988) Structure and function of voltage sensitive ion channels. *Science*, **242**, 50–61.

Cooper, J.R., Bloom, F.E. and Roth, R. (1988) *The Biochemical Basis of Neuropharmacology*. Oxford University Press, Oxford.

Godfraind, T. and Govoni, S. (1989) Increasing complexity revealed in regulation of Ca^{2+} antagonist receptor. *TIPS*, **10**, No. 8.

Iversen, L.L. and Goodman, E. (Eds) (1986) *Fast and Slow Signalling in the Nervous System*. Oxford University Press, Oxford.

Rang, H.P. and Dale, M.M. (1987) *Pharmacology*. Churchill Livingstone, Edinburgh.

Rogers, H.J., Spector, R.G. and Trounce, J.R. (1986) *Textbook of Clinical Pharmacology*. Hodder & Stoughton, London.

Snyder, S.H. and Bennet, J.P. (1976) Neurotransmitter receptors in the brain: Biochemical identification. *Ann. Rev. Physiol.*, **38**, 153–75.

Spedding, M. (1987) Three types of Ca^{2+} channel explain discrepancies. *TIPS*, **8**, No. 4.

Starke, K., Gothert, M. and Kilbinger, H. (1989) Modulation of neurotransmitter release by presynaptic autoreceptors. *Physiol. Rev.*, **69**, 865–989.

Trends in Pharmacological Science (1987) Vol. 8, issue 12 : A Feast of Short Reviews Celebrating 100 issues of TIPS.

ibid. January, 1991. Receptor nomenclature supplement (to be updated annually).

2 *Acetylcholine*

Acetylcholine was the first identified neurotransmitter; it was acetylcholine which was released in response to stimulation of the vagus nerve in Loewi's experiment described in Chapter 1. Studies have shown that acetylcholine is a neurotransmitter at the following sites:

1. All preganglionic nerve terminals (both parasympathetic and sympathetic) of the autonomic nervous system (ANS), i.e. autonomic ganglia;
2. All postganglionic parasympathetic nerve terminals;
3. The neuromuscular junction (of voluntary nerve to skeletal muscle);
4. The adrenal medulla;
5. The central nervous system (CNS);
6. Postganglionic sympathetic nerve terminals at sweat glands.

Drugs which imitate or mimic the actions of acetylcholine are called cholinomimetics, while drugs which prevent the actions of acetylcholine are called anticholinergics.

2.1 SYNTHESIS

Acetylcholine (ACh) is synthesized in cholinergic nerve terminals by the enzyme choline acetyltransferase, which acetylates choline, the acetyl group being transferred from acetyl coenzyme A (Fig. 2.1).

Choline is a dietary constituent which is carried in plasma and taken up into cholinergic neurones by a high affinity uptake system. Choline derived from the hydrolysis of acetylcholine is also taken up by this system.

2.1.1 Control of acetylcholine synthesis

Choline acetyltransferase (CAT) is found in the cytoplasm and appears not to be saturated by the concentrations of choline normally found

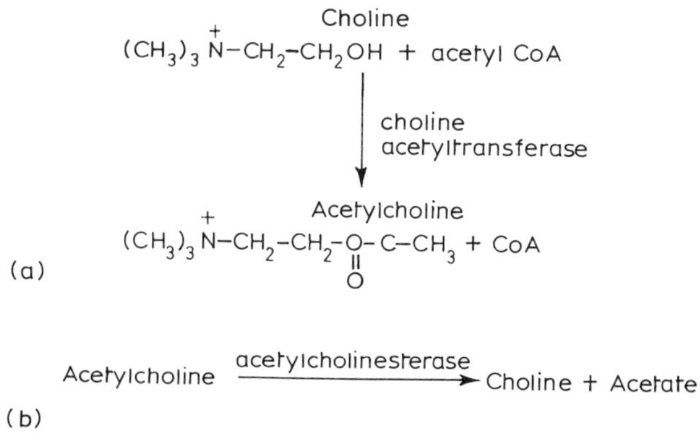

Choline

$$(CH_3)_3 \overset{+}{N}-CH_2-CH_2OH \; + \; acetyl \; CoA$$

$$\downarrow \quad choline \\ acetyltransferase$$

Acetylcholine

$$(CH_3)_3 \overset{+}{N}-CH_2-CH_2-O-\underset{\underset{O}{\|}}{C}-CH_3 \; + \; CoA$$

(a)

$$Acetylcholine \xrightarrow{\;\; acetylcholinesterase \;\;} Choline \; + \; Acetate$$

(b)

Figure 2.1 Acetyl choline: (a) synthesis; (b) breakdown.

within nerve terminals; thus, increasing the concentration of choline available to the enzyme increases the rate of ACh synthesis. The availability of choline within the cholinergic neurone determines the availability of ACh for release, thus drugs such as hemicholinium and triethylcholine which compete with choline for uptake into nerves, will reduce the synthesis and hence the availability of ACh; they are used only experimentally.

2.2 STORAGE

ACh is stored in nerve terminals in subcellular structures called vesicles, together with ATP and a soluble protein called vesiculin. A small proportion of the total ACh is free in the cytoplasm; possibly this is *en route* from the site of synthesis to the vesicles, or from storage to sites of release. As discussed below, there is not complete consensus as to the function of the vesicles within cholinergic – or indeed any other – nerve terminals; the suggestion that neurotransmitter is released from vesicles is tentative. No clinically used compounds which can selectively alter ACh storage are known. Compounds which affect ACh synthesis (as discussed above) can cause changes in the amount of ACh stored within nerves, but this action is indirect.

2.3 RELEASE

It is a widely held view that the ACh stored within vesicles is released by exocytosis in response to depolarization of the nerve terminal. In common with other neurotransmitter-release mechanisms, calcium is necessary for ACh release. Experiments show that newly synthesized ACh is released in preference to that which is stored.

When micro-electrodes have been inserted into skeletal muscles in the region of motor end-plates (a postsynaptic site) miniature depolarizations have been recorded. An action potential arriving at the presynaptic nerve terminal releases enough ACh to produce depolarization of the motor end-plate region of sufficient magnitude to initiate an action potential within the muscle. It has been suggested that the miniature depolarization (miniature end-plate potential, mepp) occurs as a result of the spontaneous release of ACh from a single vesicle, and that end-plate potentials arise from the simultaneous release of the contents of many vesicles.

2.3.1 Modification of release

Botulinum toxin prevents ACh release from all cholinergic nerves. While no mechanism for this activity is clearly established, the most likely explanation appears to be that botulinum toxin interferes with calcium ions, which must enter the terminal to initiate ACh release. Whether calcium entry or calcium-linked exocytosis is inhibited is not known. Black-widow-spider venom is reputed to cause an explosive release of ACh from cholinergic nerves.

2.4 CHOLINERGIC RECEPTORS

On the basis of experimental evidence, two main types of cholinergic receptor are recognized (Fig. 2.2). The classification is based on the actions of two drugs of plant origin which were found to mimic the effects of stimulating specific nerves; it was later established that the nerves release ACh.

1. Muscarine was found to mimic the effect of parasympathetic-nerve stimulation, and the receptors on neuroeffector tissues with a parasympathetic nerve supply are known as muscarinic receptors.
2. At the neuromuscular junction of voluntary nerves and skeletal muscle, the effects of nerve stimulation can be mimicked by nicotine; these are therefore called nicotinic receptors. Nicotine mimics ACh at ganglia in the ANS, at the adrenal medulla and in parts of the CNS.

2.4.1 Muscarinic receptors in the ANS

Muscarinic-receptor stimulation occurs physiologically when the parasympathetic nervous system is active during rest and sleep.

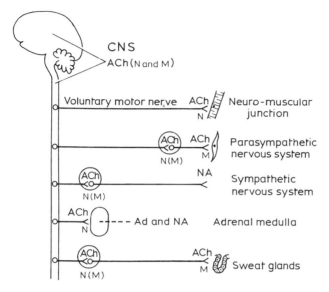

Figure 2.2 Cholinergic synapses in the nervous system. N, nicotinic receptors; M, muscarinic receptors; Ad, adrenaline; NA, noradrenaline.

(a) The heart

Muscarinic receptors mediate a slowing in the rate of contraction of the heart and a decrease in the force of contraction. The major sites of action are in the atria, the sino-auricular node and the sino-ventricular node. Excess muscarinic-receptor stimulation in the heart can lead to cardiac arrhythmias, atrio-ventricular block and eventually cardiac arrest.

(b) Blood vessels

Muscarinic receptors are found on the endothelial cells which line the walls of blood vessels. Activation of the muscarinic receptors leads to synthesis of nitric oxide (NO) which diffuses from the endothelial cells into the smooth muscle of the blood vessel. NO is a potent vasodilator which was originally called EDRF (endothelium derived relaxation factor) when it was found that the ability of blood vessels to relax in response to stimulation of muscarinic receptors was abolished when the endothelial lining of the blood vessels was stripped away. Most blood vessels do not have a parasympathetic innervation.

(c) Exocrine gland secretion

Activation of muscarinic cholinergic receptors leads to increased exocrine secretion in salivary glands, mucus glands in the bronchi, parietal cells in the stomach, enzyme-secreting cells in the intestine and sweat glands in the skin. The sweat glands respond to cholinergic stimuli, even though anatomically they are sympathetically innervated.

(d) Gastro-intestinal smooth muscle

This responds to muscarinic-receptor stimulation by an increase in tone and an increase in contractions and peristalsis. Intestinal cramps and pain may occur, as well as involuntary defaecation.

(e) Ureters and bladder

Both are contracted by muscarinic-receptor stimulation. This may lead to involuntary micturition.

(f) The eye

There are three responses in the eye resulting from muscarinic-receptor activation: (1) the ciliary muscles contract, thus causing relaxation of the lens, which is then focused for near vision; (2) the iris circular-muscle fibres contract, resulting in constriction of the pupil. This is called miosis; (3) third effect of muscarinic-receptor stimulation is the reduction of the elevated intra-ocular pressure which occurs in glaucoma. The mechanism(s) by which intra-ocular pressure is reduced may include better drainage through the canal of Schlemm, which dilates following contraction of the ciliary muscles, and vasodilation, which may also improve drainage (see Fig. 2.3).

Muscarinic receptors belong to the G protein regulated super family of cell surface receptors. Five genes have been identified which code for muscarinic receptors, all of which conform to the seven transmembrane region structure outlined in section 1.9. On the basis of pharmacological studies only three muscarinic receptor sub-types are at present recognized and they have been designated M1, M2 and M3; M2 has also been called a cardiac muscarinic receptor.

The M1 and M3 receptors are linked to G_s to phospholipase C, and their activation leads to increased synthesis of inositol trisphosphate and diacyl glycerol. The M2 receptor has two effector pathways; inhibition of adenylyl cyclase leading to decreased cyclic AMP synthesis and G protein regulated opening of potassium ion channels.

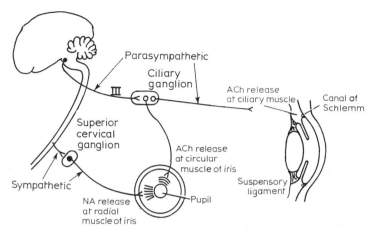

Figure 2.3 The autonomic innervation of the eye.

2.4.2 Nicotinic receptors in the ANS

ACh is a neurotransmitter at both parasympathetic and sympathetic ganglia in the ANS. Most cholinergic receptors at ANS ganglia are nicotinic receptors, because they are excited (and later blocked) by nicotine. The effects of stimulation of nicotinic receptors at autonomic ganglia will depend on the location of the ganglion, and on the branch of the autonomic nervous system to which the ganglion belongs. Stimulation of parasympathetic ganglia will lead to a parasympathomimetic response; stimulation of sympathetic ganglia will lead to a sympathomimetic response.

The adrenal medulla may be regarded as a specialized sympathetic autonomic ganglion, adapted for the release of adrenaline into the blood stream. The release of ACh from preganglionic nerve endings leads to excitation of nicotinic receptors, which initiates the release of adrenaline and noradrenaline.

2.4.3 Nicotinic receptors at the neuromuscular junction

Voluntary nerves of skeletal muscle release ACh in response to an action potential; the ACh released acts postsynaptically on nicotinic receptors to initiate muscular contraction.

Nicotinic ACh receptors belong to the ligand regulated ion channel super family of receptors. Two sub-types of nicotinic receptor are recognized, a sub-type found in skeletal muscle and a sub-type found in nerves. The structure of the skeletal muscle nicotinic receptor is described in Figs 1.2 and 2.4; the structure of the neuronal nicotinic receptor is still incompletely defined. Nicotinic ACh receptors regulate the movement of sodium, potassium and calcium ions into and out of cells. The two sub-types of nicotinic receptor can be sub-divided on the basis of their pharmacological properties (sections 2.4.8 and 2.6.4).

2.4.4 Cholinergic receptors in the CNS

Many pharmacological, physiological and biochemical studies indicate that there are both muscarinic and nicotinic receptors in the CNS.

Cholinergic mechanisms have been associated with the following central nervous system functions.

1. Drugs which activate or inhibit cholinergic mechanisms in the brain have varying effects on arousal and wakefulness in man and animals.
2. Cholinergic mechanisms are involved in nausea, vomiting and possibly vertigo. Cholinomimetics are able to induce all of the above symptoms; muscarinic receptor blockers can have anti-emetic action.
3. Resting tremor and rigidity are associated with excess cholinergic activity in Parkinson's disease (section 4.6.1). Muscarinic receptor blockers were the first (and continue to be) effective treatments for these symptoms of Parkinson's disease.
4. Drugs which block muscarinic receptors are known to inhibit short term memory. Loss of CAT immuno reactivity of cholinergic neurones in senile and other forms of dementia has led to the suggestion that cholinergic mechanisms are critically involved in memory processes. These observations have been tenuously linked with research strategies for improving cognitive defects, to date with rather disappointing results. This probably arises from lack of suitable pharmacological agents rather than improper formulation of the cholinergic hypotheses in cognition and memory. Experiments in animals indicate a critical role for the septohippocampal pathway for deposition of memory; reversal of this memory deficit by direct injection of cholinomimetic compounds or by implantation of acetylcholine secreting cells into senile rats, or into animals where lesions of the septohippocampal pathway have been made, and restoration of ability to learn, suggest an important role for acetylcholine in these processes.

2.4.5 Muscarinic receptor agonists

ACh is rarely used, except experimentally, because it is rapidly broken down by the enzyme acetylcholinesterase. **Muscarine** is only of toxicological and historical interest. It is found in some fungi (e.g. *Amanita muscaria*) and contributes to some forms of mushroom poisoning. It is not used therapeutically.

Methacholine is not such a good substrate for acetylcholinesterase as is ACh; it is administered subcutaneously and has a short duration of action. The major effects are on the heart, blood vessels and intestines, and on the bladder by action on muscarinic cholinergic receptors. The effects of methacholine are potentiated by acetylcholinesterase inhibitors.

Carbachol and **bethanechol** are used for their effects at muscarinic receptors primarily in the gut and the bladder. **Carbachol** and **bethanechol** are not destroyed by acetylcholinesterase, and have a longer duration of action than **methacholine**. They can be given either orally or by subcutaneous injection.

Pilocarpine and **arecoline** are muscarinic-receptor agonists extracted from plants. Their use is restricted to application to the eye.

Oxotremorine is used for its muscarinic actions in the central nervous system. Selective agonists for sub-types of muscarinic receptor are not available.

2.4.6 Muscarinic receptor blockers (antimuscarinic agents)

Atropine and **hyoscine (scopolamine)** block the effects of ACh at all muscarinic receptors. The effect of atropine-like drugs in the ANS is qualitatively the opposite to the effects of ACh at muscarinic sites; the final effect of blocking muscarinic receptors in any tissue, however, will depend on the degree of sympathetic tone.

Following the blocking of muscarinic receptors in the eye, the ciliary muscles are relaxed, and there is loss of ability to focus on near objects. The circular muscles in the iris relax, and the pupil is dilated. Intra-ocular pressure may rise, resulting in glaucoma. Because vagal effects are decreased, increased heart rate occurs following atropine treatment. Salivary, bronchial and gastro-intestinal secretions are reduced by atropine. Decreased salivary and bronchial secretion is a desirable feature of atropine in pre-anaesthesia medication. The tone and peristalsis of gastro-intestinal smooth muscle is reduced by atropine-like compounds. Atropine inhibits sweat-gland secretion by blocking muscarinic receptors associated with cholinergic post-ganglionic sympathetic nerves. Antimuscarinic drugs have actions within the central nervous system. They are effective in controlling tremor and rigidity in Parkinson's disease, and although atropine was the first compound used for this action, **benztropine**, **benzhexol** and **orphenadrine** are now more commonly used (section 4.6.1). They are also effective anti-emetics, especially in the treatment of travel sickness. The anti-emetic activity of antihistamine drugs (H_1 receptor blockers) appears to be related to their muscarine-receptor blocking actions. **Ipratropium** is a muscarinic blocker delivered by aerosol and inhaled, used in the treatment of bronchial asthma.

The actions of muscarinic receptor blockers outlined above can be attributed to their actions at M2 and M3 muscarinic receptor sites. The muscarinic receptor regulating acid secretion in the stomach is an M1 muscarinic receptor which can be competitively blocked by **pirenzepine**. This is a selective muscarinic M1 receptor blocker which at therapeutic doses does not block the muscarinic receptors in other parts of the body. It is used in the treatment of gastric and duodenal ulceration caused by excess acid secretion.

Before the advent of the selective M1 muscarinic receptor blocker **pirenzepine**, non-selective M1 and M2 receptor blockers such as **hyoscine** and **atropine** were ineffective in the treatment of excess gastric acid secretion; the doses of these compounds necessary to achieve suppression of gastric acid secretion were too high and resulted in unacceptable adverse effects including blurred vision, dry mouth, constipation and inability to urinate.

2.4.7 Nicotinic-receptor agonists

At autonomic ganglia, the adrenal medulla, and at the neuromuscular junction, **nicotine** initially acts as an agonist but, following initial excitation, it acts as a depolarizing nicotinic-receptor blocker. It is not used therapeutically, but is found in some insecticides and in tobacco. Dimethyl phenyl piperazinium (DMPP) is used experimentally to stimulate nicotinic receptors in ganglia.

2.4.8 Nicotinic-receptor blockers (antinicotinic agents)

Hexamethonium is a competitive inhibitor, and **mecamylamine** and **pempidine** appear to be mixed (i.e. competitive and non-competitive) blockers of nicotinic receptors at autonomic ganglia. They are sometimes used as antihypertensive agents but, as they are not selective for sympathetic ganglia and cause many side effects, they are rarely used.

(+)-**Tubocurarine** acts as a competitive blocker of ACh at nicotinic receptors at the neuromuscular junction and is a skeletal muscle relaxant. If the concentration of ACh at the motor end-plate is increased (by administration of an anticholinesterase such as **neostigmine**), then the (+)-**tubocurarine** block of nicotinic receptors can be reversed. (+)-**Tubocurarine** is not completely selective for nicotinic receptors at the skeletal–neuromuscular junction; it also blocks the nicotinic receptors in autonomic ganglia, and so in use it can have ganglion-blocking actions, which are manifested as hypotension. Additionally, (+)-**tubocurarine** can cause histamine release, which might further contribute to the hypotensive response. As (+)-**tubocurarine** is a quaternary ammonium compound, it is not highly lipid-soluble; therefore it cannot be administered orally, and is administered intravenously. It does not enter the CNS.

Gallamine and **pancuronium** are competitive skeletal muscle relaxants which have relatively weaker ganglion-blocking and histamine-releasing activity than does (+)-**tubocurarine**.

Succinylcholine (**suxamethonium**) transiently stimulates nicotinic receptors, causing depolarization of the muscle membrane; this is seen as fasciculation or twitching of the muscle. Repolarization of the muscle membrane does not subsequently occur in the presence of **succinylcholine**, and thus the muscle stays depolarized and insensitive to further stimulation by either ACh or **succinylcholine**. This type of muscular-relaxant action is called depolarizing block.

At high doses **suxamethonium** can act as an antagonist at autonomic ganglia; at these doses it has marked histamine releasing actions. Its actions are terminated by hydrolysis by plasma cholinesterases.

Suxamethonium can cause slowing of the heart due to a direct muscarinic receptor activity. This action can be reversed by **atropine**.

Atracurium is a competitive inhibitor of nicotinic receptors of the skeletal–neuromuscular junction. This molecule is stable at acid pH but at physiological pH (for example in plasma) it spontaneously hydrolyses into two inactive compounds. It thus has a short duration of action which is terminated by a mechanism which is independent of either metabolism or renal excretion.

At the nicotinic receptor in skeletal muscle and at autonomic ganglia, transmission can be blocked in one of three ways.

(a) Competitive receptor blockers

Competitive receptor blockers such as **tubocurarine** prevent acetylcholine occupying the ligand binding site responsible for opening of the cationic channel (Fig. 2.4). The action of **tubocurarine** is called competitive because increasing the concentration of acetylcholine at the site results in displacement of the **tubocurarine** and reversal of the block. This process has some characteristics of events which can be described in terms of the law of mass action, when two molecules are competing for the same active site. Competitive antagonists act as non-depolarizing blockers. **Tubocurarine** is weakly selective for skeletal muscle nicotinic receptors.

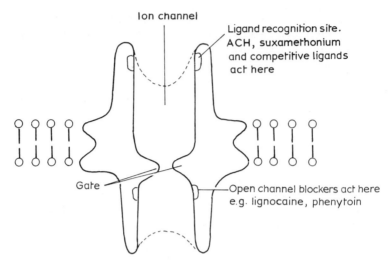

Figure 2.4 Sites at which drugs can modify the activity of nicotinic receptors. **Tubocurarine** and **atracurium** compete with ACh for the ligand recognition site and prevent opening of the gate.

(b) Membrane stabilizers; local anaesthetics; use dependent ion channel blockers

Certain drugs have the ability to block both voltage regulated and ligand regulated ion channels. They can only do so when the channels have been opened in response to a depolarizing stimulus. It is believed that the process of opening the ion channels allows the drug molecules to enter the channel and become lodged in it; once lodged in the channel the drug molecules prevent further movement of ions in the channel. This is a property shared by many drugs and it is called local anaesthetic activity or membrane stabilizing activity. Because the drug cannot lodge itself in the open ion channel until that channel has been activated, such drugs are referred to as use dependent channel blockers or open channel blockers. Drugs which are used as local anaesthetics (for example during dental surgery, e.g. **lignocaine**), and anti-epileptic drugs such as **phenytoin**, are use dependent open channel blockers (Fig. 2.4).

(c) Depolarizing blockers

This type of action appears to be confined to nicotinic receptors although it may also apply to excitatory amino acids. Depolarizing blockers such as **suxamethonium** and **decamethonium** are agonists at the ligand regulated nicotinic receptor of skeletal muscle and ganglia. High concentrations of acetylcholine such as those which occur following treatment with anticholinesterase can also cause depolarization block. The depolarization which occurs, is due to the persistent occupation of nicotinic receptors resulting in the inactivation of voltage regulated sodium ion channels. The depolarization induced by the agonists appears to inactivate the potential sensor on adjacent voltage regulated channels.

Nicotinic receptors at ganglia mediate fast transmission by operation at ligand regulated ion channels. Ganglionic transmission can be modified by a wide range of chemical messengers. Most of these effects are mediated by activation of G protein regulated mechanisms; for example M1 muscarinic receptors mediate a slow hyperpolarization at ganglia while LHRH (luteinizing hormone releasing hormone) causes a slow depolarization at ganglia. By such interactions, subtle regulation of ganglionic transmission may be achieved.

2.4.9 Cholinergic autoreceptors

Inhibitory muscarinic autoreceptors have been described on somatodendritic sites in both the autonomic and central nervous systems

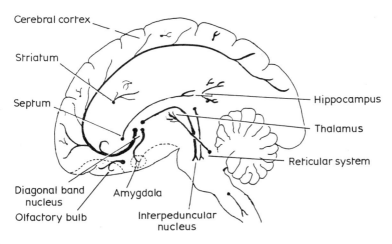

Figure 2.5 Diagrammatic representation of acetylcholine pathways in the human brain. Mapping of the cholinergic pathways in the CNS has been achieved by a combination of immunocytochemical and biochemical measures. Acetylcholine using pathways have been traced using CAT immunoreactivity and binding of muscarinic and nicotinic ligands to suitably prepared sections of brain has been used to map the distribution of receptors. There is generally good agreement between different mapping techniques.

and axon terminal muscarinic autoreceptors have also been described at both sites. It has been suggested that M1 and M3 receptors are located predominantly postsynaptically, and that M2 receptors are predominantly autoreceptors on axon terminals and are therefore presynaptic.

2.5 INACTIVATION OF ACETYLCHOLINE

Following release from cholinergic nerves, ACh is broken down by the action of enzymes. In the synaptic cleft, on both pre- and postsynaptic structures, ACh is hydrolysed by the enzyme acetylcholinesterase (AChE). Acetylcholinesterase hydrolyses ACh to choline and acetate (Fig. 2.1). The choline can then be taken back up into the presynaptic neurone, and there it is used to synthesize new ACh. Acetylcholinesterase is only found in close association with cholinergic neurones. Plasma cholinesterase (also known as pseudocholinesterase, non-specific cholinesterase and butyrylcholinesterase) is found in plasma and liver, as well as in nervous tissue, but it is not associated with sites of cholinergic transmission. Whereas the function of acetylcholinesterase is to terminate the activity of released ACh, the functions of plasma pseudo-cholinesterase are presumably to inactivate any ACh which might escape destruction by AChE, and to inactivate any other choline esters which might find their way around the body (for example **succinylcholine**).

2.5.1 Inhibitors of acetylcholinesterase

Drugs which either reversibly or irreversibly inhibit the enzyme acetylcholinesterase, and thus delay the destruction of released ACh, are called anticholinesterases. Anticholinesterases will inhibit acetylcholinesterase at any site to which they can gain access, and they will mimic and potentiate the effects of ACh at all parasympathetically innervated structures, at nicotinic receptors in autonomic ganglia and skeletal muscle, and in the central nervous system.

Physostigmine (eserine) is an anticholinesterase whose effect in man lasts for a few hours. Owing to its high lipid solubility, it can be given orally, and it enters the CNS. **Pyridostigmine** or **neostigmine** may be given orally (but usually intravenously), and because of low lipid solubility do not enter the CNS; actions are more pronounced at the skeletal–neuromuscular junction than at autonomic sites. They are used in the treatment of myasthenia gravis.

Organophosphate anticholinesterases constitute a class of long-lasting (irreversible) compounds. These compounds have been mainly used as insecticides and as agents of warfare. **Diisopropyl fluorophosphate (DFP or dyflos)** sarin and soman are examples of this group of compounds.

Plasma pseudocholinesterases are also inhibited by most anticholinesterases. **Succinylcholine** is probably the major substance which is destroyed by the action of plasma pseudocholinesterase. In individuals deficient in this enzyme, the effects of **succinylcholine** will be prolonged.

2.6 THERAPEUTIC APPLICATIONS AND CONSEQUENCES OF DRUGS ACTING AT CHOLINERGIC SYNAPSES

2.6.1 Synthesis of acetylcholine

2.6.2 Storage of acetylcholine

Alteration of ACh synthesis by the inhibition of choline acetyltransferase has not been found to be therapeutically useful. Similarly, decreasing the availability of choline in neurones by preventing its uptake by **hemicholinium** or **triethylcholine** has not been found to have therapeutic applications. Increasing cholinergic activity by the administration of choline in the diet, either as choline itself or in lecithin, is currently being evaluated in some degenerative conditions of the CNS involving cholinergic neurones.

Patients with presenile dementia (Alzheimer's disease) have been found, on biopsy and autopsy, to have reduced ACh-synthesising capacity, especially in the hippocampus, because the levels of choline acetyltransferase are found to be significantly lower than in the matched controls. Since one of the symptoms in these conditions is loss of short-term memory, and **hyoscine** is known to inhibit short-term memory, it has been suggested that cholinergic mechanisms are of importance in this process. Studies in patients with Alzheimer's disease and in normal volunteers have shown that increasing cholinergic function by giving choline or lecithin, or inhibiting AChE with **physostigmine** does not improve memory or the other symptoms of the disease.

2.6.3 Drugs which affect acetylcholine release

Black-widow-spider venom is reported to cause an explosive release of ACh from synaptic vesicles. The symptoms of poisoning are those of general stimulation at all cholinergic sites, followed by depolarization block at all nicotinic sites. The mechanism involved in the explosive release is not understood.

In contrast to the symptoms seen with substances which increase the amount of ACh in the synaptic cleft, food poisoning with botulinum toxin results in symptoms of combined muscarinic and nicotinic receptor block at all cholinergic sites. Botulinum toxin irreversibly blocks ACh release from all cholinergic neurones. Death occurs as a result of paralysis of the respiratory muscles.

2.6.4 Drugs acting at cholinergic receptors

(a) Cholinergic agonists at muscarinic receptors

ACh is not used therapeutically, except as a solution for application to the eye. **Methacholine, carbachol** and **bethanechol** can be used for their actions at muscarinic receptors. Their use is generally restricted to application to the eye to produce relaxation of the lens, constriction of the pupil and lowering of intra-ocular pressure in glaucoma. When given orally or subcutaneously (**carbachol** and **bethanechol**) or subcutaneously (**methacholine**), they all stimulate contractions of the intestines and of the bladder; they are used in the treatment of paralytic ileus and atony of the bladder. These substances must not be given intravenously as they are liable to cause cardiac arrest.

Pilocarpine is used for its parasympathomimetic actions when directly applied to the eye.

(b) Cholinergic blockers at muscarinic receptors

Cholinergic muscarinic receptor blockers have been used for many conditions where attenuation of parasympathetic activity has been indicated. At present the following therapeutic use is made of these compounds.

Hyoscine with a perhaps fortuitous combination of CNS and ANS actions, has been a favoured compound in pre-anaesthetic medication, since it combines sedative, amnesic, anti-emetic, antivagal and antisecretory activity.

Atropine has some CNS-stimulant actions causing agitation and excitement; its actions in the ANS are similar to those of **hyoscine**, and it is used in pre-anaesthetic medication. **Atropine** is used in emergency treatment of sinus bradycardia: it blocks vagal tone, and increases heart rate.

Prior to the introduction of **pirenzepine** (a selective M2 muscarinic blocker), non-selective muscarinic receptor blockers were ineffective in the treatment of excess gastric acid secretion. At doses needed to reduce acid secretion, they caused unacceptable adverse effects. **Pirenzepine** is used to reduce acid secretion in the stomach (see also H2 blockers, section 6.6.4).

Ipatropium is a muscarinic receptor blocker which is available as an inhalant in the treatment of bronchospasm associated with asthma. The route of administration reduces systemic adverse effects. **Atropine** and other antimuscarinic drugs (**propantheline** and **mebeverine**) have been used to reduce excess intestinal motility associated with colic; this inevitably leads to constipation and other adverse effects associated with blockade of muscarinic receptors.

Propanthelene and **terodiline** are used in the treatment of urinary frequency, as they help stabilize detrusor muscle irritability.

In ophthalmology, antimuscarinic agents are used topically by instillation into the eye. They cause dilation of the pupil, which is useful when examination of the retina is desired. Relaxation of the lens is important in the measurement of refractive errors. **Atropine** and **hyoscine** have a duration of action of several days when applied to the eye, and attempts to limit their duration of action by application of agonists (**pilocarpine** or an anticholinesterase) are usually only partly successful. **Tropicamide** and **homatropine** have a much shorter duration of action when applied to the eye (about 8 h) and thus offer an advantage over other agents if only a short duration of action is required. These agents must be used with care in persons with narrow-angle glaucoma, in whom they can precipitate a dangerous rise in intra-ocular pressure.

Muscarinic receptor blockers are sometimes used in the treatment of tremor and rigidity which occurs in Parkinson's disease. Agents with other actions (e.g. **benztropine** and **benzhexol**, section 4.6.1) have been introduced to replace **hyoscine** and **atropine**, and they are frequently administered in combination with **L-DOPA**, which is of most value in overcoming the inability to initiate movement in Parkinson's disease. The inhibition of salivary gland secretion by these compounds can usefully reduce drooling of saliva, but can cause discomfort due to dry mouth. Blurred vision, constipation and urinary retention will also result from the use of these compounds.

Atropine is also of value in the treatment of some forms of mushroom poisoning where the toxic agent is a muscarine-like compound. It is of value in the treatment of poisoning with anticholinesterases, but its effects are limited to antagonism of the muscarinic effects of ACh in the ANS and CNS, whereas nicotinic actions, especially at the skeletal–neuromuscular junction, are not affected.

Anticholinergic drugs which enter the CNS have some anti-emetic actions. This property is also observed with some H_1-histamine receptor blockers (section 6.6.4) and, since these compounds have muscarinic-receptor-blocking properties, they probably owe their anti-emetic actions to their anticholinergic effects.

Poisoning following ingestion of deadly nightshade (*Atropa belladona*), which contains **atropine**, or **henbane** (*Hyoscyamus niger*), which contains **hyoscine**, causes symptoms associated with the inhibition of muscarinic receptors. The CNS effects of the drugs differ: there is excitation and restlessness following **atropine**, and drowsiness and sedation following **hyoscine**. At toxic doses, both compounds cause delirium and coma. Anticholinesterases (for example, **physostigmine**) can be used in the treatment of anticholinergic poisoning. Numerous other compounds can block muscarinic receptors, and thus give rise to the symptoms of cholinergic block, such as blurred vision, dry mouth and constipation. Among the more common compounds which have anticholinergic action of this kind are the tricyclic antidepressants (**imipramine, desipramine** and **amitriptyline**; section 3.6.5), some H_1-receptor blockers (**mepyramine** and **promethazine**; section 6.6.4) and some neuroleptics (**chlorpromazine** and **perphenazine**; section 4.6.4).

(c) Cholinergic drugs at nicotinic receptors at ganglia

Substances which selectively stimulate nicotinic receptors at ganglia have not found therapeutic applications. Anticholinesterases will facilitate cholinergic transmission at this site, as well as at all other cholinergic sites. Drugs which block cholinergic nicotinic receptors at ganglia (ganglion blockers) include **hexamethonium, mecamylamine** and **trimetaphan**; these are very rarely used.

Ganglion blockers lower blood pressure by decreasing the sympathetic tone to noradrenergic nerve endings, thus decreasing vasoconstriction in arteriolar beds. Ganglion blockers interfere with baroreceptor reflexes, which is particularly noticeable when people stand up from sitting or lying down. Because reflex vasoconstriction is inhibited, pooling of blood occurs in the large veins of the abdomen and in the legs, and insufficient venous return to the heart can cause dizziness and fainting. This is known as orthostatic or postural hypotension. As ganglion blockers do not act selectively at only sympathetic ganglia, block of transmission in parasympathetic ganglia leads to constipation due to decreased tone and motility of the gut; dry mouth; urinary retention and difficulty of micturition; and blurring of vision due to inability to focus.

(d) Cholinergic drugs at nicotinic receptors at the skeletal–neuromuscular junction

Competitive and depolarizing skeletal-muscle relaxants demand separate consideration.

Competitive skeletal-muscle relaxants. (+)-**Tubocurarine, gallamine,**

pancuronium and **atracurium** are used to obtain relaxation of skeletal muscle during surgery, during manipulation of fractured limbs or dislocated joints and occasionally in some convulsive disorders. Since the muscles of respiration (intercostal and diaphragm) are paralysed, it is important to maintain adequate artificial ventilation. Neuromuscular blocking agents have no anaesthetic or analgesic action. A suitable degree of anaesthesia and analgesia must be maintained, because the person cannot respond to painful stimuli.

They compete with ACh for nicotinic receptors on skeletal muscle; their duration of action can be decreased and terminated by anti-cholinesterases, which increase the ACh available in the synaptic cleft. When anticholinesterases are used for this purpose, an antimuscarinic drug is usually given to prevent excess parasympathomimetic stimulation due to the action of the anticholinesterase at parasympathetic nerve endings. Actions of these compounds at sites other than the skeletal–neuromuscular junction are discussed in section 2.4.8.

Depolarizing skeletal-muscle relaxant. **Succinylcholine** (**suxamethonium**), has a nicotine-like action at the skeletal–neuromuscular junction, causing first some contractures or fasciculations of the muscle (which might account for the tenderness reported following their use), and then a depolarizing block which results in muscular relaxation. **Succinylcholine** has a brief duration of action, since it is rapidly destroyed by plasma pseudocholinesterase. It is thus of value in situations where a brief period of muscular relaxation is needed, as in tracheal intubation and electroconvulsive treatment (ECT). Patients with low plasma pseudocholinesterase activity will have a drastically prolonged response to **succinylcholine**. It is not possible to reverse the neuromuscular relaxation following **succinylcholine** by treatment with an anti-cholinesterase, because the increased levels of ACh which occur will reinforce the depolarization block and not reverse it (section 2.4.8).

2.6.5 Drugs which inhibit acetylcholinesterase

Anticholinesterases are used (1) for their actions on the eye (physostigmine applied topically); (2) to activate the gastro-intestinal and urinary systems; (3) to reverse the actions of competitive muscle relaxants at the skeletal–neuromuscular junction; (4) to reverse the central actions of muscarinic receptor blockers (including the anticholinergic effects of high doses of tricyclic antidepressants); and (5) in the treatment of myasthenia gravis.

Myasthenia gravis is a condition characterized by weakness of the muscles, and inability to do muscular work. The causes of this condition have only recently been made clear; previously it was assumed that there

Table 2.1 Summary of drugs which modify cholinergic transmission

Mechanism	Drug	Effect	Uses
Synthesis	**Choline**	Increased synthesis	Experimental
	Hemicholinium	Decreased choline uptake	Experimental
Storage	—	—	—
Release	Black-widow-spider venom	Explosive release of ACh	Toxicological and experimental
	Botulinum toxin	Blocks release of ACh	
Receptors	**ACh**	Nicotinic - receptor activation	Experimental
	Nicotine		
	TMA		
	Hexamethonium	Nicotinic receptor block at ganglia	Antihypertensives
	Mecamylamine		
	Pempidine		
	(+)-Tubocurarine	Nicotinic receptor block at neuromuscular junction (NMJ)	Skeletal muscle relaxation
	Gallamine		
	Pancuronium		
	Atracurium		
	Suxamethonium	Desensitizing (depolarizing) block at NMJ	Skeletal muscle relaxation

Drug	Action	Clinical use
ACh **Muscarine** **Methacholine**	Muscarinic receptor activation	Parasympathetic stimulation
Carbachol **Bethanechol** **Pilocarpine**	Non-selective agonists	
Atropine **Hyoscine**	Muscarinic receptor block Non-selective blockers	Anti-emetic Pre-operative medication AChE poisoning Parkinson's disease Bronchospasm Antacid
Benztropine **Ipratropium** **Pirenzepine** **Telenzepine**	Selective M1 receptor blockers	
AFDX 116 **Methoctramine**	Selective M2 receptor blockers	
Hexahydrosiladifenol	Selective M3 receptor blocker	
Physostigmine **Neostigmine** **Pyridostigmine**	Reversible anticholinesterases	Reversal of competitive NMJ block Myasthenia gravis
Inactivation of metabolism **Dyflos (DFP)**	Irreversible anticholinesterase	

was a defect in either the synthesis, storage or release of ACh from motoneurones. At present it is generally agreed that myasthenia gravis is an auto-immune response in which the patient produces antibodies to his or her own nicotinic cholinergic receptors in skeletal muscle. The antibodies are released into the circulation and combine with the 'antigenic' receptors on the motor end-plate, and thus inactivate them. Thymectomy and treatment with corticosteroids is found to be of value, presumably in the first case by destroying the site of antibody production, and in the second by reducing the production of antibodies. Anticholinesterases find a major use in this condition, for both diagnosis and therapy.

Edrophonium is a very short-acting anticholinesterase, and is used for the diagnosis of myasthenia gravis. A short period of improved muscular function occurs, and this constitutes a confirmatory diagnosis. Anticholinesterases such as **neostigmine** and **pyridostigmine** which can be taken orally and are relatively free of CNS side effects (as they do not cross the blood–brain barrier), are used therapeutically. Acetylcholinesterase at sites other than those at the skeletal–neuromuscular junction is also inhibited during treatment with **neostigmine** or **pyridostigmine** and thus patients must routinely take antimuscarinic drugs (usually **propanthalene**) to prevent autonomic cholinomimetic side effects.

Myasthenia gravis is a condition which varies in intensity from time to time, and the dosage of anticholinesterase has to be matched to the degree of disability. It is important to distinguish the effects of inadequate anticholinesterase medication or worsening of the condition (myasthenic crisis) from those of excessive anticholinesterase medication, leading to a depolarizing block (section 2.4.8) as a result of excess accumulation of ACh in the synapse (cholinergic crisis). It is important that antimuscarinic drugs are not used to excess, as the appearance of excessive parasympathomimetic activity can give warning of the approaching cholinergic crisis. If these effects are masked by excess muscarinic receptor block, then identification of the type of crisis is made difficult. Symptoms of a myasthenic crisis should improve with extra anticholinesterase treatment (**edrophonium** can be helpful in differential diagnosis), while in cholinergic crisis symptoms remain unchanged or become worse. In a cholinergic crisis it is usually necessary to withdraw all anticholinesterase treatment, to be prepared to give artificial ventilation and to give adequate antimuscarinic drugs to prevent excess parasympathetic activity on the heart.

Patients with myasthenia gravis are very sensitive to drugs such as **tubocurarine, succinylcholine** and **aminoglycoside** antibiotics which have some curare-like action, local anaesthetics (including β-adrenoceptor blockers), benzodiazepines and hypokalaemic agents (digitalis and diuretics), all of which can exacerbate the symptoms.

Poisoning with anticholinesterases. Such poisoning can occur in situations associated with the manufacture of anticholinesterases or their use as insecticides, and overdose can occur during treatment of myasthenia gravis. Both muscarinic and nicotinic actions of ACh are enhanced in systemic poisoning; depolarizing block of the skeletal muscles of respiration, arrest of the heart due to excess vagomimetic actions and effects on the central nervous system all contribute eventually to death. The muscarinic effects are treated with adequate doses of antimuscarinic agents such as **atropine**.

It is possible to reactivate acetylcholinesterase with agents such as **pralidoxime**, which displace the anticholinesterase from the enzyme by combining with the anticholinesterase to produce a stable compound which is not strongly bound to the active site of the enzyme. To be effective, treatment with **pralidoxime** must be started as soon as possible after exposure to the anticholinesterase.

The major drugs which modify cholinergic transmission are summarized in Table 2.1.

FURTHER READING

Barnes, P.J., Minnette, P. and Maclagan, J. (1988) Muscarinic receptor subtypes in airways. *TIPS*, 9, No. 11.

Bowen, D.M. and Davison, A.N. (1980) Biochemistry of Alzheimers disease, in *The Biochemistry of Psychiatric Disturbances* (ed. G. Curzon), John Wiley, Chichester and New York, pp. 113–28.

Changeux, J-P., Giraudat, J. and Dennis, M. (1987) The nicotinic acetylcholine receptor: molecular architecture of a ligand-regulated ion channel. *TIPS*, 8, No. 12.

Cuello, A.C. and Sofroniev, M.V. (1984) The anatomy of CNS cholinergic neurons. *TINS*, 74–8.

Heilbronn, E. and Stalberg, E. (1978) The pathogenesis of Myasthenia Gravis. *Ann. Rev. Pharmacol.*, 20, 337–62.

Moncada, S., Herman, A.G. and Vanhoutte, P. (1987) Endothelium-derived relaxing factor is identified as nitric oxide. *TIPS*, 8, No. 10.

3 *Noradrenaline*

Noradrenaline (norepinephrine, norarternol) is a catecholamine neurotransmitter in postganglionic sympathetic nerves and within the central nervous system (CNS). Noradrenaline is released from the adrenal medulla. The term 'catecholamine' refers to the chemical classification in which substances with two adjacent hydroxyl (OH) groups on a benzene ring are said to have a 'catechol' nucleus.

Whereas noradrenaline was eventually identified as being the neurotransmitter of postganglionic sympathetic nerves, the activity of adrenaline was studied in early work, and the term 'adrenergic' is frequently used when the noradrenergic system is being discussed. The term 'noradrenergic' will be used throughout this text whenever possible but, where the conventions necessitate the use of the term 'adrenergic', then the reader must bear in mind that, with a few exceptions, the prefix 'nor' is understood.

3.1 SYNTHESIS

Noradrenaline (NA) is made from the aromatic amino acid L-tyrosine, which is hydroxylated to L-3,4-dihydroxyphenylalanine (L-DOPA) by the enzyme tyrosine hydroxylase (Fig. 3.1). L-Tyrosine is actively taken up into noradrenergic nerve terminals. Tyrosine hydroxylase is located in the cytoplasm of noradrenergic neurones, and is the rate-limiting enzyme in NA synthesis. The enzyme requires reduced pteridine cofactor, molecular oxygen and ferrous ions for activity. L-DOPA is decarboxylated in the cytoplasm to dopamine by L-aromatic amino acid decarboxylase (DOPA-decarboxylase). The dopamine formed is actively taken up into granular noradrenaline storage vesicles in which the dopamine is hydroxylated (on the β-carbon atom) to form noradrenaline by the enzyme dopamine-β-hydroxylase, which is found associated with noradrenaline storage vesicles. Dopamine-β-hydroxylase is a copper-

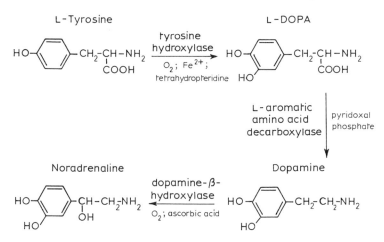

Figure 3.1 Synthesis of noradrenaline.

containing enzyme which requires molecular oxygen and uses ascorbic acid as a cofactor.

In the adrenal medulla, and in some neurones in the CNS, noradrenaline is converted into adrenaline by methylation of the amine group by the enzyme phenylethanolamine-N-methyl transferase (PNMT).

3.1.1 Control of noradrenaline synthesis

Tyrosine is not an essential amino acid, since it can be synthesized by hydroxylation of phenylalanine. The rate of synthesis of noradrenaline cannot be usefully altered by varying either dietary tyrosine or phenylalanine intake. The activity of tyrosine hydroxylase, and thus the rate of synthesis of noradrenaline, is under the influence of the following factors.

1. Increased concentration of noradrenaline within nerve terminals decreases the rate of conversion of tyrosine into L-DOPA as a result of 'end-product' inhibition of tyrosine hydroxylase by noradrenaline.
2. The activity of tyrosine hydroxylase is influenced by sympathetic activity. Increased action potential traffic to nerve terminals leads to phosphorylation of tyrosine hydroxylase; this increases the activity of tyrosine hydroxylase. Decreased neuronal activity leads to dephosphorylation of tyrosine hydroxylase and decreased rate of noradrenaline synthesis.

3.1.2 Inhibitors of the enzymes of NA synthesis

α-Methyl-*p*-tyrosine is used experimentally to inhibit tyrosine hydroxylase, and thus to prevent the synthesis of all catecholamines (noradrenaline, dopamine and adrenaline). Inhibitors of aromatic amino acid decarboxylase in tissues outside the CNS (e.g. carbidopa) which are used with L-DOPA in the therapy of Parkinson's disease seem not to alter noradrenergic function significantly.

Dopamine-β-hydroxylase is most efficiently inhibited by copper-chelating agents such as **diethyldithiocarbamate**, **disulfiram** and FLA63. Disulfiram is used in the treatment of alcoholism and also experimentally.

3.1.3 α-Methyl-DOPA, α-methyl-noradrenaline as a false transmitter

Criteria for a true neurotransmitter are considered in Chapter 1. If a structural analogue of the neurotransmitter precursor enters the neurone, a structural analogue of the transmitter may be synthesized and released. Such an analogue is known, on release from the neurone, as a **false transmitter**. The best example of this class of compound is α-methyl-noradrenaline (α-methyl-NA), which is synthesized from α-methyl-DOPA by the actions of aromatic amino acid decarboxylase and dopamine-β-hydroxylase within noradrenergic neurones (Fig. 3.2).

This reaction occurs in all noradrenergic neurones, both in the sympathetic nervous system and in the CNS. When used therapeutically, α-methyl-DOPA does not inhibit either the aromatic amino acid decarboxylase or dopamine-β-hydroxylase. Its antihypertensive action depends on its conversion into α-methyl-NA. The mechanisms and sites of action of α-methyl-DOPA and α-methyl-NA in hypertension are discussed in sections 3.4.4 and 3.6.1.

3.2 STORAGE

It is generally believed that noradrenaline is stored within the nerve terminal in more than one type of storage complex, and possibly at more

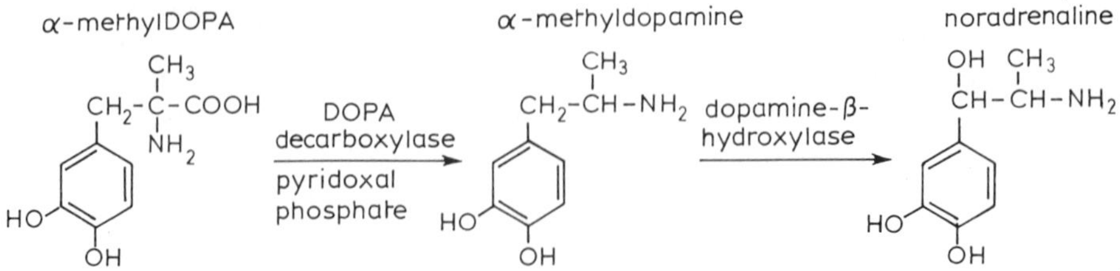

Figure 3.2 Synthesis of α-methyl noradrenaline.

than one anatomical location. The most readily described storage form of noradrenaline is a granular complex found within vesicles in noradrenergic nerve terminals. The granular complex consists of noradrenaline bound with ATP, several proteins collectively called chromogranins (which include the enzyme dopamine-β-hydroxylase) and divalent metal ions including Mg^{2+}, Ca^{2+} and Cu^{2+}. The granular complex containing noradrenaline is sometimes called the 'insoluble' or 'particulate' noradrenaline storage fraction, and this apparently vesicular storage form has been further subdivided by some authorities. There is also evidence to suggest that noradrenaline might be found in a 'soluble' or 'free' storage fraction within the cytoplasm of the nerve terminal. This noradrenaline fraction might have a role in controlling the rate of noradrenaline synthesis, by affecting the activity of tyrosine hydroxylase (which is the rate-limiting step in synthesis and which is found in the cytoplasm; section 3.1.1). How such a soluble storage fraction of noradrenaline might be protected from the actions of MAO is not clear.

3.2.1 Uptake of noradrenaline and dopamine into storage vesicles

Dopamine is taken up from the cytoplasm into noradrenaline storage vesicles, where, by the action of dopamine-β-hydroxylase, it is converted into noradrenaline (Fig. 3.3). The uptake of dopamine and noradrenaline into storage vesicles is an active-transport process which requires ATP as an energy source and Mg^{2+} to activate the ATPase enzyme (Mg^{2+}-dependent ATPase). It is important to realize that this Mg^{2+}-dependent uptake process of noradrenaline and dopamine into storage vesicles is a different and separate process from the neuronal-uptake process for noradrenaline across the nerve-cell membrane (section 3.5.1), which is a Na^+-dependent-ATPase process (uptake 1).

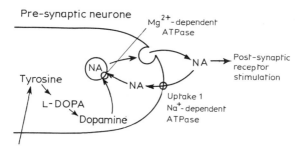

Figure 3.3 Uptake sites for noradrenaline in the sympathetic nerve ending. Similar mechanisms operate in dopamine and 5-hydroxytryptamine neurones.

3.2.2 Disruption of storage

The stability of the noradrenaline–ATP–protein–metal ion storage complex can be disrupted by **reserpine** and **tetrabenazine** (Rauwolfia alkaloids), which chelate Mg^{2+}. Noradrenaline and dopamine uptake into noradrenergic storage vesicles is reduced, and the storage complex is disrupted. As noradrenaline can no longer be retained within the storage vesicles, it leaks into the cytoplasm, where it is destroyed by MAO (section 3.5.2). Following **reserpine** or **tetrabenazine**, the amount of noradrenaline found within nerves and available for release is severely reduced, and consequently noradrenergic transmission is impaired. **Reserpine** appears to have an irreversible action; normal activity of noradrenergic neurones is not re-established until new storage vesicles become available by means of axoplasmic flow from the nerve-cell body.

In the CNS, dopamine and 5-hydroxytryptamine are stored in vesicles which have many properties in common with noradrenaline storage vesicles. They share the sensitivity to **reserpine**, which, in addition to causing disruption of storage and depletion of noradrenaline, also disrupts and depletes stores of dopamine and 5-hydroxytryptamine.

Enhanced storage of noradrenaline

The storage capacity of vesicles appears to be greater than the amount of noradrenaline normally found within noradrenergic nerves, as noradrenaline levels within nerves can be increased by preventing noradrenaline breakdown with MAO inhibitors (section 3.6.5).

3.3 RELEASE

Experimental evidence suggests that release of noradrenaline from nerve terminals occurs by a process of exocytosis, during which the vesicular membrane fuses with the plasma membrane, and the vesicular contents – consisting of noradrenaline, ATP, dopamine-β-hydroxylase and chromogranins – are released into the synaptic cleft. Calcium influx is essential for noradrenaline release.

3.3.1 Regulation of noradrenaline release

Evidence for the existence of autoreceptors originated from observations in experiments where the release of radioactive noradrenaline was measured from sympathetic nerve endings. In the presence of α_2-adrenoceptor blockers, release was increased; in the presence of α_2-

agonists, the release was decreased. The observations have subsequently been replicated and confirmed for many other neurotransmitter systems. Whether autoreceptors have any physiological role in regulation of transmitter release is not firmly established; they are likely to remain the target for pharmacological investigations for a considerable time.

The α_2-autoreceptor inhibits release of noradrenaline from noradrenergic nerve endings. The α_2-adrenoceptor is linked by an inhibitory G protein to α adenylyl cyclase; activation of this receptor decreases activity of adenylyl cyclase and decreases synthesis of cyclic AMP. It is suggested that cyclic AMP regulates the activity of a calcium calmodulin sensitive protein kinase which in turn modulates the availability of noradrenaline containing synaptic vesicles for release.

There is evidence for the existence of presynaptic β-adrenoceptors on some noradrenergic nerve endings. β-adrenoceptors are linked by a stimulatory G protein to adenylyl cyclase; activation of these receptors leads to increased synthesis of cyclic AMP. Presynaptic β-adrenoceptors would be facilitatory in modulating release of noradrenaline.

3.3.2 Inhibition of noradrenaline release

Guanethidine, debrisoquine, bretylium and **bethanidine** are anti-hypertensive agents, which decrease the amount of noradrenaline released in response to sympathetic-nerve stimulation. They decrease the effectiveness of sympathetic nerve stimulation and are collectively known as adrenergic neurone blockers. The mechanisms by which they prevent noradrenaline release have not been fully established, and individual agents might prove to act differently. Suggestions include a local anaesthetic effect, which stabilizes the vesicular membrane, thus making exocytosis more difficult, or an interference with the entry of calcium into the nerve terminal, thus inhibiting stimulus secretion coupling. In order to act, adrenergic neurone blockers have to be taken up into the nerve terminal by the neuronal-uptake process (uptake 1; section 3.5.1). During this uptake process, guanethidine exhibits some noradrenaline-displacing activity, which can cause a rise in blood pressure if it is given by rapid intravenous injection.

At physiological pH, adrenergic neurone blockers are only sparingly lipid soluble; thus they do not cross the blood–brain barrier and their actions are restricted to the peripheral nervous system.

3.3.3 Displacement of noradrenaline from storage sites by drugs

Indirectly acting sympathomimetic amines are substances which displace noradrenaline from storage vesicles and result in its release into the synaptic cleft. This release process is not dependent on the entry of calcium. **Amphetamines, tyramine, ephedrine** and **phenylpropranolamine** are indirectly acting sympathomimetic amines which are so called because they mimic the effects of sympathetic stimulation. Their mechanisms of action are complex and varied and include the following.

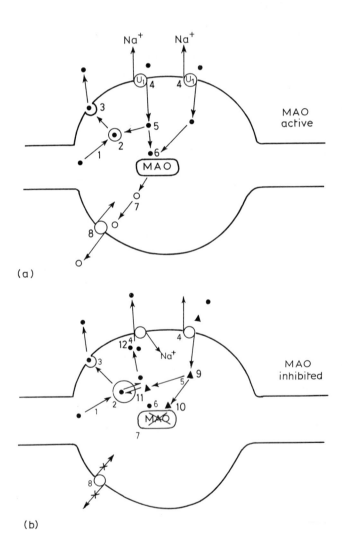

(a)

(b)

1. Uptake by and inhibition of uptake 1.
2. As they are all lipid soluble, they enter the nerve terminal by passive diffusion.
3. They are either substrates for, or reversible inhibitors of, MAO.
4. They are able to displace noradrenaline from the newly synthesized vesicular store.

These drugs accumulate in noradrenergic nerve terminals and displace noradrenaline into the cytoplasm. The noradrenaline does not get broken down because MAO is inhibited by the drugs. Noradrenaline leaves the nerve terminal by passive diffusion (down a concentration gradient) or by exchange diffusion processes (Fig. 3.4). Noradrenaline in the synaptic cleft stimulates postsynaptic alpha and beta adrenoreceptors thus mimicking the effects of sympathetic stimulation.

3.4 ADRENOCEPTORS

When NA is released from the presynaptic nerve terminal, it crosses the synaptic cleft, and initiates a response in the postsynaptic tissue by

---

Figure 3.4 The mechanisms of action of indirectly acting sympathomimetic amines, in this example, (+)-**amphetamine**. This model applies to noradrenaline, dopamine and 5-HT secreting neurones: the figure depicts a single axonal varicosity. The normal sequence is shown in (a). 1, synthesis; 2, vesicular uptake; 3, exocytotic release in response to an action potential and Ca^{2+} entry; 4, uptake of amine by the high affinity uptake 1 system. This requires energy (from ATP) and Na^+ is carried out of the cell in exchange for an amine molecule; 5, the amine is now either taken back into the storage vesicle (2); or 6, inactivated by oxidative deamination by MAO; the deaminated acid metabolites (7), are removed from the cell by an active transport pump (8) which has affinity for weak acids.

The changes which occur in the presence of (+)-**amphetamine** are shown in (b). Processes 1, 2 and 3 proceed as normal. 4, **amphetamine** enters the cell by passive diffusion (because it is lipid soluble), and by (ii) active uptake. Because it is taken up, **amphetamine** is a competitive inhibitor of amine uptake. Within the cell, process 5 is substituted by 9: **amphetamine** has a high affinity for MAO and acts as a competitive inhibitor of this enzyme (10), so it prevents deamination of the amine (processes 6 and 7 do not operate). **Amphetamine** also enters the storage vesicles (11), where it exchanges for the amine molecule, which is displaced into the cytoplasm. A high concentration of amine can build up within the cell, so much so, that the normal exchange of amine for Na^+ is driven in reverse (12); amine is transported out and Na^+ is carried into the cell. High concentrations of amine can occur in the synaptic cleft and lead to activation of post synaptic receptors. ▲: (+)− amphetamine; ●: noradrenaline; ○: acid metabolites.

combining with one of two main types of receptor called α-adrenoceptors and β-adrenoceptors. Whereas noradrenaline is the transmitter at most sites within the postganglionic sympathetic nervous system, the type of receptor found postsynaptically to noradrenergic nerves depends on the tissue.

Adrenoceptors have been classified into two major sub-types designated α and β initially on the basis of studies using agonists, and more recently by means of selective agonists and competitive blockers, by characterization of effector/transducer mechanisms, and by gene sequencing. All adrenoceptors appear to belong to the G protein regulated receptor super family.

All α-adrenoceptors belong to the G protein regulated super family. Activation of the receptors leads to transduction by different effector systems; activation of α_1-adrenoceptors leads to increased IP3 and diacyl glycerol synthesis via phospholipase C activation. Activation of α_2-adrenoceptors leads to inhibition of adenylyl cyclase or opening of a G protein regulated potassium ion channel.

α-Adrenoceptors are characterized by being most sensitive to the actions of **noradrenaline, phenylephrine** and **adrenaline** and least sensitive to the actions of **isoprenaline**. Furthermore, low concentrations of **phentolamine** and **phenoxybenzamine** can competitively and selectively block the responses to **noradrenaline** and **phenylephrine** but not those to **isoprenaline**.

β-Adrenoceptors are characterized by being most sensitive to the actions of **isoprenaline** and least sensitive to the actions of **noradrenaline** and **phenylephrine**. At low concentrations, **propranolol** can competitively and selectively block the responses to **isoprenaline**, but not those to **phenylephrine**.

On the basis of studies made with substances which selectively activate β-adrenoceptors, a further subclassification of β-adrenoceptors has been made: adrenoceptors in the heart are called β_1-adrenoceptors, and β-adrenoceptors found anywhere else outside the CNS are called β_2-adrenoceptors.

α-Adrenoceptors can be divided into two groups on the basis of sensitivity to agonists and selectivity of antagonists. α_1-Adenoreceptors are found exclusively postsynaptically and they are more sensitive to the actions of **adrenaline** and **noradrenaline** than to the actions of **clonidine** and/or α-**methyl noradrenaline**. **Prazosin** is a selective α_1-adrenoceptor blocker. α_2-Adrenoceptors include presynaptic autoreceptors but α_2-adrenoceptors are also found at postsynaptic sites in the autonomic nervous system and in the CNS. **Yohimbine** is a selective inhibitor of α_2-adrenoceptors.

3.4.1 Postsynaptic α-adrenoceptors in the ANS

(a) The eye

α-Adrenoceptors mediate contraction of the radial muscle of the iris, thus leading to pupillary dilation (mydriasis). α-Adrenoceptors are activated to keep the eyelid open; inhibition may lead to ptosis (drooping of the upper eyelid).

(b) Blood vessels

α-Adrenoceptor stimulation in most blood vessels leads to vaso-constriction as a result of contraction of smooth muscle arranged in a circular manner.

(c) Gastro-intestinal tract

Non-sphincteric smooth muscle of the gastro-intestinal tract relaxes to sympathetic stimulation, the relaxation being mediated by both α- and β-adrenoceptors. Sphincter muscles, however, generally contract in response to sympathetic stimulation of α-adrenoceptors.

(d) Urinogenital system

Seminal-vesicle and vas-deferens smooth muscle contract following stimulation of α-adrenoceptors. Constriction of the trigone and sphincter in the bladder is mediated by activation of α-adrenoceptors. The smooth muscle of the uterus responds variably to α-adrenoceptor stimulation, depending on the stage of the oestrus cycle and on pregnancy.

α-Adrenoceptors mediate contraction of the smooth muscle of the pilomotor muscles, the nictitating membrane, the splenic capsule and the salivary glands.

3.4.2 Postsynaptic β-adrenoceptors in the ANS

β-Adrenoceptors are all linked to adenylyl cyclase by a G_s regulatory protein; activation of β-adrenoceptors leads to increased synthesis of c-AMP.

(a) The heart

β_1-Adrenoceptors are found in the heart, and stimulating them results in an increase in both the force and rate of contraction. β_1-Adrenoceptors also mediate an increase in the contraction velocity in both the atria and ventricles. This can lead to cardiac arrhythmias during excess sympathetic stimulation.

(b) Blood vessels

In skeletal muscle and in the liver, β_2-adrenoceptor stimulation mediates vasodilation by relaxation of smooth muscle.

(c) Tracheal and bronchial smooth muscle

β_2-Adrenoceptor stimulation on bronchial smooth muscle leads to relaxation, causing bronchodilation.

(d) Gastro-intestinal tract

Only non-sphincteric smooth muscle appears to have β_2-adrenoceptors in the gastro-intestinal tract; in common with actions at α-adrenoceptors, sympathetic stimulation at β-adrenoceptors leads to relaxation and decrease in motility and tone.

(e) Urinogenital system

β_2-Adrenoceptors mediate relaxation of the uterus in pregnancy. The detrusor muscle of the bladder is relaxed during sympathetic stimulation of β-adrenoceptors.

(f) Renin release

Renin release from the juxtaglomerular cells of the arterioles in the kidney is under sympathetic influences acting on β-adrenoceptors.

(g) Metabolic actions

Activation of β-adrenoceptors leads to enhanced glycogenolysis in liver, increased lipolysis in adipose tissue and decreased release of insulin from the pancreas. The increased glycogenolysis leads to elevated blood glucose and decreased glucose uptake into skeletal muscle as a result of decreased insulin release. This leaves glucose available for metabolism by the brain. The increased lipolysis provides a ready energy source in the form of fatty acids for skeletal muscle. In situations of stress, therefore, the metabolic actions of catecholamines acting on β-adrenoceptors serve to ensure a diversion of energy sources to brain and skeletal muscle. The existence and functions of proposed β_3-adrenoceptors in brown adipose fat, where they may regulate thermogenesis and deposition of fat, is being investigated.

3.4.3 Adrenoceptors and noradrenaline in the CNS

The distribution of noradrenaline in the CNS has been established by histochemical, biochemical and bio-assay techniques. Most noradrenergic nerve-cell bodies are in the brainstem, and at least four pathways to other parts of the CNS originate from this region (Fig. 3.5).

(a) NA in the lower brainstem

There is a dense noradrenergic innervation in the medulla oblongata, consisting of neurones with short axons. The nucleus of the tractus solitarius and also the dorsal vagal nucleus have major noradrenergic innervations. It is believed that the noradrenergic neurones in this region are concerned with central control of peripheral sympathetic tone, and are thus a major site of central control of blood pressure. Experimental evidence suggests that both α- and β-adrenoceptors are found here.

(b) Spinal pathways

Also arising in the medulla oblongata, there are three descending spinal noradrenergic pathways: in the dorsal and ventral horns and in the lateral

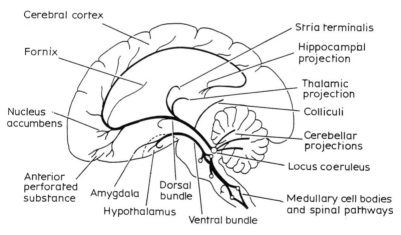

Figure 3.5 The noradrenaline innervation of the CNS. The nerve-cell bodies and dentrites are concentrated in the brainstem and locus coeruleus; the distribution of axon terminals is shown.

sympathetic column. The lateral column is associated with vasomotor control, while ventral-horn NA neurones acting on α-adrenoceptors participate in the control of flexor muscles.

(c) Dorsal NA bundle

In the pons, the locus coeruleus contains nerve-cell bodies from which axons ascend in the dorsal NA bundle to innervate the thalamus, the dorsal hypothalamus, the limbic system, the hippocampus and the neocortex.

(d) Ventral NA bundle

Arising ventrally and caudally to the locus coeruleus is another ascending NA fibre bundle, which has a more ventral pathway and terminates in the hypothalamus and in sub-cortical limbic sites. Among the suggested functions controlled by the latter two systems are mood and behaviour. Our knowledge of the control of these functions is poor, but it appears that responses to stress and depression are under noradrenergic control. Noradrenergic mechanisms are important in the sleep–waking cycle, for example it is believed that the locus coeruleus prevents motor activity during rapid eye movement (REM) sleep. It is also suggested that noradrenergic mechanisms are activated during 'rewarding' activities, for intracranial self-stimulation experiments have shown that noradrenergic processes are important in reinforcement, which is a form of positive feedback apparently important in learning. The rewarding component of food intake is believed to be under noradrenergic control, and this complements the idea that these mechanisms are part of the general reward-recognition system. Thermoregulation has also been associated with noradrenergic mechanisms centred in the hypothalamus.

3.4.4 Adrenoceptor agonists

Adrenoceptor agonists form two sub-classes. Directly-acting compounds (direct sympathomimetics, i.e. those which combine directly with the postsynaptic receptor) can be selective for either α- or β-adrenoceptors when used in appropriate doses. Indirectly-acting agonists (indirectly-acting sympathomimetics) cause the release of NA from presynaptic terminals, and thus act on whichever adrenoceptors are found postsynaptically. Indirectly-acting sympathomimetics are not selective for either α- or β-adrenoceptors.

(a) Directly acting α-adrenoceptor agonists

Noradrenaline released from sympathetic nerve terminals will act on whatever adrenoceptor is present postsynaptically. When injected intramuscularly (rarely intravenously), NA exerts its effects mainly through α-adrenoceptors (Fig. 3.6). In the cardiovascular system this is seen as vasoconstriction, leading to a rise in systolic, diastolic and mean arterial pressure; this, in turn, leads to reflex slowing of the heart through the baroreceptor reflex arc.

Noradrenaline does not enter the systemic circulation following oral intake, because it is metabolized in the gut and liver and because it is not very lipid soluble; low lipid solubility also prevents entry into the CNS if it is given parenterally.

Adrenaline initiates both α- and β-adrenoceptor-mediated responses (Fig. 3.6). In practice, most actions can be related to β-adrenoceptor stimulation.

Phenylephrine can be given orally, subcutaneously or intravenously and, as it has actions confined almost exclusively to α-adrenoceptors, its effects on the cardiovascular system are similar to those of noradrenaline. It is used as a nasal decongestant, where vasoconstriction of the nasal mucosa results in decreased secretions, thus alleviating the symptoms of

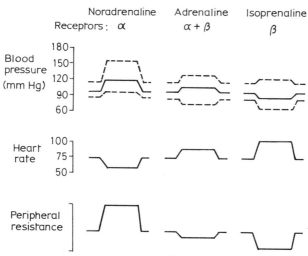

Figure 3.6 The cardiovascular effects of intravenous infusion of noradrenaline, adrenaline and isoprenaline in man. In the blood pressure trace the upper dotted line shows changes in systolic pressure, the lower dotted line changes in diastolic pressure. (After Allwood, Cobbold and Ginsburg, 1963. By permission of the Medical Department, British Council.)

the common cold. Topical application to the eye leads to pupillary dilation (mydriasis). **Metaraminol** and **methoxamine** are selective α-adrenoceptor agonists.

Clonidine is an imidazoline derivative with strong α_2-adrenoceptor-stimulant actions. If given intravenously in man, it causes a transient rise in blood pressure (due to vasoconstriction), followed by a prolonged fall in mean arterial pressure. This latter fall in blood pressure is due to the actions of **clonidine** in the CNS, where it acts as an α-adrenoceptor agonist in neurones which control peripheral sympathetic tone. The precise location of these α-adrenoceptors is not established, but the medullary region of the brain stem is important, as direct application of α-adrenoceptor agonists at this site causes decreased peripheral sympathetic tone and reduces vasoconstriction. At the same time, there is vagally-mediated slowing of the heart, resulting in decreased cardiac output. Combined, these actions of **clonidine** lead to hypotension. (**Clonidine** may also have actions at presynaptic α-receptors – section 3.3.1.)

α-**Methyl-DOPA** is metabolized to α-**methyl-noradrenaline** (section 3.1.3), which is an α_2-adrenoceptor agonist in the CNS and causes a fall in blood pressure. α-**Methyl-noradrenaline** is nearly equipotent with noradrenaline as an α-adrenoceptor or β-adrenoceptor agonist in the peripheral nervous system; thus it is not likely that any α-**methyl-noradrenaline** synthesized within peripheral sympathetic nerves would contribute to the hypotensive action. α-**Methyl-DOPA** does not inhibit DOPA decarboxylase *in vivo*. The hypotensive actions of α-**methyl-DOPA** are believed to be centrally mediated.

(b) Directly-acting β-adrenoceptor agonists

When released from sympathetic nerves, noradrenaline will act at whichever adrenoceptors are found postsynaptically. Physiologically, therefore, NA acts as an agonist at all β-adrenoceptors. When injected parenterally, the α_1-adrenoceptor stimulant action of NA is most pronounced and, unless all α-adrenoceptors are blocked, the β-adreno-ceptor-mediated actions of NA are not manifested.

Injected parenterally, **adrenaline** has mainly β-adrenoceptor-mediated effects. In the cardiovascular system, adrenaline causes a fall in diastolic pressure due to vasodilation of skeletal muscle beds. Systolic pressure increases as a result of the increased cardiac output due to increased cardiac rate and stroke volume. There is a small rise or no change in mean arterial pressure. The increase in force and rate of contraction of the heart is due to the direct action of adrenaline on cardiac β-adrenoceptors (Fig. 3.6). Higher doses of adrenaline can cause a rise in systolic, diastolic and mean arterial pressure; in such circumstances, vasoconstriction due to α-adrenoceptor stimulation overcomes the vasodilation following β-adrenoceptor activation. As they have a lower threshold of activation, β-adrenoceptors appear to be more sensitive to adrenaline than are α-adrenoceptors. **Adrenaline** has been used as a bronchodilator.

Isoprenaline (isoproterenol) is a synthetic catecholamine. In the doses used in most experimental and therapeutic situations, **isoprenaline** acts only at β-adrenoceptors, and is free of significant α-adrenoceptor stimulant action. **Noradrenaline, adrenaline** and **isoprenaline** are unselective, affecting both β_1- and β_2-adrenoceptors. In the cardio-vascular system, **isoprenaline** causes vasodilation and a fall in total peripheral resistance, leading to a fall in diastolic pressure. The force and rate of cardiac contractions are increased, leading to increased cardiac output and a rise in systolic pressure. Pulse pressure is increased, but mean arterial pressure either falls slightly or is unchanged (see Fig. 3.6). **Isoprenaline** is a more potent β-adrenoceptor agonist than are **noradrenaline** and **adrenaline**.

When inhaled as a spray or aerosol, **isoprenaline** causes bronchodilation by acting on bronchial β_2-adrenoceptors.

Salbutamol and **terbutaline**, when used in appropriate doses, act as selective β_2-adrenoceptor agonists. They are used as bronchodilators in the treatment of asthma.

Dobutamine and **xameterol** are selective β_1-adrenoceptor agonists used to increase cardiac output in heart failure and some forms of shock.

(c) Indirectly-acting adrenoceptor agonists

Amphetamine and other indirectly-acting sympathomimetics release NA from all sympathetic nerves, and therefore they are not selective in their actions at α- and β-adrenoceptors. These compounds can also block the re-uptake of NA into nerve terminals and so prolong the action of released **noradrenaline**. In the cardiovascular system, indirectly-acting sympathomimetic amines cause a rise in blood pressure, and reflex cardiac slowing may occur despite the activation of cardiac β_1-adrenoceptors. **Amphetamine** has stimulant actions in the CNS, where it increases arousal and both mental and motor activity. These responses are probably brought about by actions on noradrenergic and dopaminergic neurones (section 4.6.3). As in the sympathetic nervous system, **amphetamine** mediates the release of transmitters in the CNS, and also blocks their re-uptake.

Tyramine is an indirectly-acting sympathomimetic amine found in cheese and other dairy products (especially those prepared by fermentation), and in alcoholic drinks. Normally, **tyramine** is completely inactivated by MAO when taken in the diet. If MAO is inhibited by drugs (section 3.5.2), then the sympathomimetic activity of tyramine may be observed as a hypertensive response. **Ephedrine** and **phenylpropanolamine** are indirectly-acting sympathomimetics which are less potent than **amphetamine**, and have less marked actions on the CNS. They can be used as vasopressor agents, but are usually used in nasal drops and in some common-cold 'cures' as nasal decongestants. Tachyphylaxis commonly occurs with indirectly-acting sympathomimetics.

3.4.5 Adrenoceptor blockers

The actions of directly and indirectly acting sympathomimetic agents can be competitively antagonized by selective adrenoceptor blockers. The availability of such compounds indicated that more than one type of adrenoceptor might exist, and they have subsequently found use in both experimental and therapeutic work.

(a) α-Adrenoceptor blockers

Phentolamine is a competitive non-selective α-adrenoceptor blocker which, when given intravenously, has a short duration of action. In the cardiovascular system, it causes a fall in blood pressure due to vasodilation, in part as a result of the blocking of α-adrenoceptors, and in part as a result of the direct action of **phentolamine** on blood vessels. The decrease in blood pressure can cause reflex speeding of the heart. **Phentolamine** is used in the diagnosis of phaeochromocytoma (hypertrophy of the adrenal medulla which results in high circulating levels of catecholamines, causing hypertension), and in the emergency treatment of hypertensive crisis. **Phenoxybenzamine** is an irreversible blocker of α_1 and α_2-adrenoceptors used in the treatment of phaeochromocytoma.

 Prazosin is a competitive α_1-adrenoceptor blocker (at postsynaptic receptors), especially in vascular smooth muscle. It has hypotensive activity.

 Ergotamine is a partial agonist at α-adrenoceptors, initially acting as an agonist, later as a receptor blocker. Its mechanism of action in the treatment of migraine is not established.

 Labetolol has both α- and β-adrenoceptor-blocking activity. It is used as a hypotensive agent.

(b) β-Adrenoceptor blockers

β-Adrenoceptor blockers are used as anti-arrhythmic agents, as antihypertensive drugs, in the treatment of angina, prophylaxis of migraine, and in the control of skeletal-muscle tremor. In addition to being competitive β-adrenoceptor blockers, many agents in this class have membrane-stabilizing activity (local anaesthetic action).

(−)-**Propranolol** (**1-propranolol**) is the isomer of propranolol with significant β-adrenoceptor-blocking properties. (+)-**Propranolol**, while retaining local anaesthetic activity, does not have significant β-adrenoceptor-blocking actions.

β-Adrenoceptor blockers were originally developed for potential use in the therapy of cardiac disease, especially angina. They are effective in this condition, since they reduce sympathetic drive to the heart and therefore reduce the work done by the heart. An unexpected and important action of β-adrenoceptor blockers is their ability to lower blood pressure, especially when this is elevated above the normal range. Several mechanisms have been suggested for the antihypertensive actions of β-blockers; it is not possible to state categorically which (if any) of the proposed explanations is correct. It is likely that, in different circumstances, each proposed mechanism assumes a more (or less) significant role, or that a combination of actions is necessary. Suggested mechanisms include the following.

1. β-Adrenoceptor blockers reduce the response of the heart to sympathetic stimulation; decreased heart rate and force of contraction lead to lowering of cardiac output. As blood pressure is related to the product of cardiac output and peripheral resistance, lowered cardiac output might be expected to decrease blood pressure. This does not occur, however, since peripheral resistance increases reflexly in response to decreased cardiac output. It is unlikely, therefore, that this mechanism alone accounts for the hypotensive response.
2. Renin is produced by the juxtaglomerular cells of the kidney. There is a sympathetic nervous supply to these cells, and released noradrenaline acts on β-adrenoceptors at this site. Renin is an enzyme, which speeds the conversion of angiotensinogen (a plasma globulin protein) into angiotensin I, a peptide consisting of 10 amino acid residues (see also section 9.1). Angiotensin I is further broken down by peptidyl peptidase enzymes into angiotensin II and angiotensin III, most of this second breakdown occurring in the lungs. Inhibition of the angiotensin converting enzyme (ACE, peptidyl peptidase angiotensin I to angiotensin II converting enzyme), by drugs such as **captopril** and **elanapril**, leads to reduced circulating angiotensin II. These drugs have proved useful in the treatment of hypertension and heart failure.

Angiotensin II is an 8 amino acid residue peptide which has three actions which can result in a rise of blood pressure. First, it is one of the most potent directly acting vasoconstrictor substances known. Second, it facilitates noradrenaline release from sympathetic nerve endings. Third, angiotensin II stimulates aldosterone secretion and release from the adrenal cortex. Aldosterone facilitates sodium ion retention in the kidney tubules, and thus passive retention of water. This can lead to an increase in blood volume and an increase in blood pressure. β-Adrenoceptor blockers can decrease renin release, and this will bring about a reduction in the production of angiotensin (Fig. 3.7).

3. Presynaptic β-receptors appear to be involved in the control of noradrenaline release. At low frequencies of stimulation, the release of noradrenaline is maintained by the action of noradrenaline on presynaptic β-receptors. It can be suggested, therefore, that in some forms of hypertension the presynaptic β-receptors are overactive, and maintain an excessive basal noradrenaline release. In such situations, β-receptor blockers acting at presynaptic β-receptors would tend to reverse this process, and tend to restore basal noradrenaline release to normal.

4. Experiments have been made which suggest that stimulation of β-adrenoceptors in the brainstem can raise blood pressure by increasing

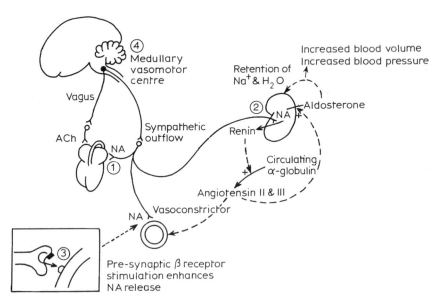

Figure 3.7 Sites of action of β-blockers. 1, block of cardiac β-adrenoceptors; 2, block of release of renin from the kidney; 3, block of presynaptic β-adrenoceptors; 4, actions in the CNS.

peripheral sympathetic tone. Competitive blockade of central β-adrenoceptors decreases peripheral sympathetic tone, and leads to lowering of the blood pressure.

Timolol is a β-blocker which is applied topically to the eye to treat glaucoma; it inhibits secretion of aqueous humour and results in a lowering of intra-occular pressure.

Atenolol and **metroprolol** are selective β_1-adrenoceptor blockers, which means that in controlled studies, they block β_1-adrenoceptors at lower concentration than that needed to also block β_2-adrenoceptors. The therapeutic advantage of selective β-adrenoceptor blockers is not firmly established as the degree of selectivity at the doses used in man is small.

3.4.6 β-Adrenoceptors and skeletal-muscle tremor

There is evidence to suggest that β-adrenoceptors in both CNS and skeletal muscle can mediate tremor, which is particularly noticeable in the arms and hands. The mechanisms and sites of action of β-adrenoceptor-mediated tremor are not well understood. β-Adrenoceptor blockers are effective in reducing tremor in Parkinson's disease, excessive essential tremor and familiar tremor. β-Adrenoceptor agonists (for example, **salbutamol** or **isoprenaline**) can cause tremor.

3.5 INACTIVATION OF NORADRENALINE

The actions of noradrenaline in the synaptic cleft are terminated by removal from the synaptic cleft by an uptake system found on presynaptic nerve endings.

3.5.1 Neuronal uptake of noradrenaline. Uptake 1

This neuronal re-uptake process is energy-dependent, requiring ATP which is broken down by a Na^+-dependent ATPase. This is a high-affinity process, which means that it is very efficient at removing low concentrations of noradrenaline from the synaptic cleft. The neuronal-uptake system transports noradrenaline into the nerve terminal. Once inside the terminal, most of the noradrenaline is further taken up into storage vesicles. This intraneuronal uptake process has been discussed (section 3.2.1 and Fig. 3.3).

6-Hydroxydopamine is a neurotoxin which is selectively taken up into catecholaminergic neurones. It then causes degeneration of those neurones, and thus it is used experimentally to make selective lesions in neuronal systems which use catecholamines as transmitters.

(a) Inhibitors of noradrenaline uptake

Included among these compounds are **cocaine** and the tricyclic antidepressants, for example, **imipramine, desipramine, amitriptyline** and **nortriptyline**. The indirectly-acting sympathomimetics (e.g. **amphetamine** and **tyramine**) also inhibit noradrenaline re-uptake, but they differ from the substances listed above in that they also significantly release noradrenaline from nerve terminals.

Observations using **cocaine** were important for the development of the idea that re-uptake of amines might be an important mechanism for terminating their actions. **Cocaine** was found to potentiate the actions of noradrenaline (a directly-acting sympathomimetic amine), but to prevent the actions of **tyramine** (an indirectly-acting amine). **Cocaine** is believed to be an inhibitor of uptake 1, and this action is independent of its local anaesthetic activity. In the CNS, cocaine has stimulant effects on mood and behaviour, actions which have been attributed to inhibition of neuronal uptake of noradrenaline and dopamine.

3.5.2 Uptake of noradrenaline into non-neuronal tissues. Uptake 2

When very high levels of noradrenaline occur in plasma – for example, after stimulation of the adrenal medulla, or intravenous injection of a catecholamine – then a significant proportion of catecholamine will be removed by uptake into non-neuronal tissues such as liver, muscle and connective tissues. This is called uptake 2. Any adrenaline or noradrenaline taken up into non-neuronal tissue either diffuses back into the circulation or, more likely, is destroyed intracellularly by the enzymes MAO and COMT, as there are no storage vesicles in such tissues (section 3.5.3).

3.5.3 Enzymatic destruction of noradrenaline. MAO and COMT

If noradrenaline is not taken up and bound in storage vesicles, it is metabolized intracellularly to biologically inactive compounds by the enzymes monoamine oxidase (MAO) and catechol-O-methyl transferase (COMT).

(a) Monoamine oxidase

This enzyme is found in all tissues which contain mitochondria, bound to their outer membranes. It is found in nerves, muscles, brain, liver and all actively metabolizing tissues. MAO oxidatively deaminates noradrenaline to 3,4-dihydroxymandelic acid, which can then be O-methylated (by COMT) to make 3-methoxy-4-hydroxymandelic acid.

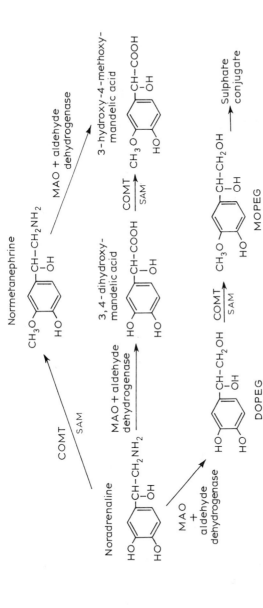

Figure 3.8 Metabolism of noradrenaline by MAO and COMT.
DOPEG, 3,4-dihydroxyphenylglycol; MOPEG, 3-methoxy-4-hydroxyphenylglycol;
SAM, 5-adenosyl-L-methionine.

This is a major urinary metabolite of noradrenaline; another pathway of metabolism yields glycols (Fig. 3.8).

Tyramine (section 3.4.4) is a substrate for MAO; when MAO is fully functional, no significant amounts of **tyramine** enter the circulation following its ingestion in food, since it is destroyed in the intestine and liver by MAO during absorption.

MAO is a convenient way of describing a group of enzymes (isoenzymes), which have different tissue distributions, substrate specificities, inhibitor characteristics and physical properties. A simple sub-classification into MAO type A and MAO type B has been suggested. MAO type A has a substrate preference for noradrenaline and 5-HT, and is selectively inhibited by **clorgyline**. MAO type B has a substrate preference for dopamine and phenylethylamine, and is selectively inhibited by **selegiline**.

(b) MAO inhibitors

These substances inhibit the enzyme and prevent deamination of substrates. Following the observation that MAO inhibitors could reverse the symptoms of depression in some patients, a great number of such compounds were made and tested, and a few are still used as antidepressants. Following inhibition of MAO, intraneuronal levels of noradrenaline (and dopamine and 5-HT) are increased in the brain, adrenal medulla and sympathetic nerves. **Iproniazid, isocarboxazid, phenelzine** and **tranylcypromine** are used as MAO inhibitors; **selegiline,** is a selective inhibitor for MAO B.

3.5.4 Catechol-O-methyl transferase (COMT)

This is an enzyme found in the cytoplasm of all tissues; high concentrations are found in the liver. COMT acts mainly on catecholamines which have been released into the circulation from the adrenal medulla and taken up into neuronal and non-neuronal tissues. COMT transfers a methyl group from S-adenosyl-L-methionine to the hydroxyl group at the three position of a catecholamine. The O-methylated metabolites of catecholamines are biologically much less active than the parent amines.

Inhibitors of COMT are being evaluated with a view to use in conjunction with **L-DOPA** in the treatment of Parkinson's disease, the rationale being that this may reduce peripheral inactivation of **L-DOPA** by O-methylation by COMT (section 4.6.1).

3.6 THERAPEUTIC APPLICATIONS AND CONSEQUENCES OF DRUGS ACTING AT NORADRENERGIC SYNAPSES

3.6.1 Drugs which affect noradrenaline synthesis. α-Methyl-DOPA. Disulfiram

α-Methyl-DOPA is a substrate for the enzymes which synthesize noradrenaline, and it is converted in noradrenergic neurones into α-methyl-dopamine (which is not a hypotensive agent) and α-methyl-noradrenaline (which is). This is released in the CNS as a false transmitter to act on α-adrenoceptors to bring about a fall in blood pressure. α-Methyl-DOPA is therefore used in the treatment of hypertension, when it is given orally.

Side effects

Sedation commonly occurs during therapy with α-methyl-DOPA, and this is probably related to its site of action in the CNS. A blocking action at some dopamine receptors is suggested by the observation that extrapyramidal side effects and hyperprolactinaemia occasionally occur (section 4.6.2). These side effects may necessitate the discontinuation of α-methyl-DOPA therapy. Postural hypotension occurs only rarely. Allergic reactions, manifested as rashes and fever, occur in a proportion of patients. As a significant amount of α-methyl-DOPA is metabolized in the liver, hepatic disease causes a potentiated response. As α-methyl-DOPA can itself cause reversible liver damage, it must not be used in patients with liver disease.

α-Methyl-p-tyrosine (AMPT) is an inhibitor of tyrosine hydroxylase which is used experimentally to inhibit synthesis of noradrenaline. It is occasionally used in the treatment of hypertension.

Disulfiram, an inhibitor of dopamine-β-hydroxylase, is not used clinically for this purpose, but it is used in the treatment of alcoholism, as it inhibits aldehyde dehydrogenase. Alcoholism is dependence on ethanol. Following initial withdrawal of ethanol, the person is given **disulfiram**. If the person drinks ethanol while still taking **disulfiram**, a build up of acetaldehyde occurs, and this leads to a severe malaise which occurs whenever ethanol is taken. This form of 'deterrent' therapy can be successful in those who can take **disulfiram** without major adverse effects.

3.6.2 Drugs which affect noradrenaline storage. Reserpine

Reserpine is occasionally used to treat hypertension, and it probably works by a combination of central and peripheral actions. Both centrally and peripherally it causes marked depletion of noradrenaline in nerve terminals. In sympathetic nerve terminals, less noradrenaline is available for release, and this probably accounts for the hypotension; such action in the CNS leads to sedation and depression, which limit the usefulness of **reserpine**.

By decreasing sympathetic tone, **reserpine** causes increased intestinal tone and motility, and this, together with the disruption of 5-HT storage, may account for the diarrhoea and colic which occur. The depression which occurs following **reserpine** can be likened to endogenous depression, and this can lead to suicidal thoughts and activities. It is not established which central neurotransmitters bring about the depression. The extrapyramidal side effects and hyperprolactinaemia which occur during therapy with **reserpine** are caused by actions on the dopamine systems (section 4.6.2).

3.6.3 Drugs which affect noradrenaline release

(a) Indirectly-acting sympathomimetics

These are sometimes used to raise blood pressure but, more usually, as nasal decongestants in some 'cures' for the common cold. In such preparations, the dose of sympathomimetic amine is usually insufficient to cause significant changes in systemic blood pressure. **Tyramine** is discussed in the context of the interactions of MAO inhibitors (section 3.6.5).

Whereas there is evidence that there is a noradrenergic component in the CNS response to **amphetamine** and related compounds (**methylamphetamine** and **phenmetrazine**), most actions of **amphetamine** on the CNS are probably dependent on the activation of dopamine systems (section 4.6.3).

(b) Adrenergic neurone blockers

Adrenergic neurone blockers, **guanethidine** and **debrisoquine** are used in the treatment of hypertension, while **bretylium** is used in the treatment of cardiac arrhythmias. They all decrease the amount of noradrenaline released from sympathetic nerve endings in response to nerve stimulation. They do not cross the blood–brain barrier, and thus are free of CNS side effects. Decrease in sympathetic function probably accounts for postural hypotension, nasal stuffiness and failure to ejaculate, which are common side effects of these compounds. Diarrhoea is a common side effect with **guanethidine**; this is probably unrelated to inhibition of sympathetic function, since it rarely occurs with other adrenergic neurone blockers.

If given rapidly by intravenous injection, **guanethidine** can cause some release of noradrenaline and thus cause a hypertensive response. This indirect sympathomimetic effect is transient, but limits the usefulness of **guanethidine** in the treatment of hypertensive crisis.

As the actions of adrenergic neurone blockers depend on the uptake of these drugs (by uptake 1) into noradrenergic nerve terminals, it is not surprising that their effectiveness as hypotensive agents is lost when indirectly acting sympathomimetic amines or tricyclic antidepressants are administered at the same time. Such compounds not only prevent the uptake of the adrenergic neurone blocker into the nerve terminal, but also displace adrenergic neurone blockers from the neuronal membrane. The hypotensive effect is lost, with a return to the hypertensive state.

Bretylium is used to control catecholamine-induced cardiac arrhythmias, and it does so by preventing the release of noradrenaline from sympathetic terminals in the heart.

3.6.4 Agonists and blockers at α- and β-adrenoceptors

(a) Direct α-adrenoceptor agonists

These are used for their vasoconstrictor actions. **Noradrenaline** is occasionally given by intravenous infusion to raise blood pressure. Any leaking around the site of entry of the infusion cannula can lead to gangrene, caused by intense vasoconstriction and lack of oxygenation. This can be prevented by infusion of **phentolamine** around the entry site of the cannula.

Constriction of blood vessels in the nasal mucosa is the basis for the use of **phenylephrine** and **oxymetazoline** in nasal drops. They are used in allergic rhinitis and the common cold, and can give relief from discomfort in middle-ear infections. Indirectly-acting sympathomimetic amines are also used for these purposes.

Noradrenaline (or more usually **adrenaline**) is used in combination with local anaesthetics to localize, potentiate and prolong the duration of anaesthesia. By constricting blood vessels at the site of injection, dispersion of local anaesthetic is delayed. **Adrenaline** in preparations of local anaesthetics is believed to be capable of causing cardiac arrhythmias in patients taking tricyclic antidepressants. It is suggested that, because uptake of adrenaline is inhibited by tricyclic compounds, adrenaline enters the circulation and reaches the heart. An alternative vasoconstrictor which does not act through adrenoceptors is the octapeptide **felypressin**. Local anaesthetics containing vasoconstrictors should not be used on fingers or toes, as the intense vasoconstriction can lead to gangrene due to decreased oxygen supply.

Adrenaline can prove life saving in anaphylactic shock or status asthmaticus, when there is hypotension, laryngeal oedema and bronchospasm; **adrenaline** increases blood pressure and relaxes bronchial muscle. In cardiac arrest, **adrenaline** may restart the asystolic heart.

The pharmacology of **clonidine** is complex, and at least two mechanisms have been proposed for its hypotensive action (section 3.4.4). It is an effective hypotensive agent which does not cause postural hypotension. At doses lower than those used to treat hypertension, **clonidine** is used in the prophylactic treatment of migraine. It presumably either reduces the responsiveness of blood vessels to circulating vaso-active compounds or, by actions on presynaptic α_2-receptors, reduces the release of noradrenaline.

Sedation is a side effect of the treatment of hypertension with **clonidine**. If the administration of **clonidine** is suddenly stopped, then a rapid and dangerous rise in blood pressure can occur. This rebound hypertension on withdrawal of **clonidine** makes it essential that patient compliance with taking the drug is good, and that the drug is not given to patients who might forget.

(b) Direct-β-adrenoceptor agonists

The major therapeutic use of direct β-adrenoceptor agonists is in the treatment of bronchoconstrictor conditions such as asthma. As **adrenaline** and **isoprenaline** are not selective for β_2-adrenoceptors, they inevitably cause tachycardia and palpitations, and can induce cardiac arrhythmias. There was a high incidence of deaths among young asthmatic patients when aerosol preparations containing high doses of **isoprenaline** were introduced. The exact cause of death was never firmly established, but cardiac toxicity is believed to have been important. While **adrenaline** is only used in aerosol form, **isoprenaline** can be given by aerosol, sublingual tablets or tablets for swallowing. The onset of action is most rapid following inhalation of aerosol, while sustained bronchodilation occurs following the swallowed tablets. Indirectly acting sympathomimetics are occasionally used to obtain bronchodilation, but the actions of compounds such as **ephedrine** are not selective for β_2-adrenoceptors, and cardiac side effects limit their usefulness.

Salbutamol, rimiterol, fenoterol and **terbutaline** are selective β_2-adrenoceptor agonists, which means that at low doses they will stimulate β_2-adrenoceptors only. Such selectivity of action offers con-siderable advantage, as it is possible to obtain useful bronchodilation without cardiac stimulation. It is important to recall, however, that selectivity is only relative, and that if the dose of these compounds is increased above the recommended therapeutic limit, palpitations, speeding of the heart, tremor and cardiac arrhythmias may occur. These compounds are available as aerosols for inhalation or tablets.

The actions of directly-acting β-adrenoceptor agonists will be potentiated by indirectly-acting sympathomimetic amines and also by **aminophylline** and **caffeine**. β-Adrenoceptor blockers will prevent β-agonists from exerting their effects, and can precipitate broncho-constriction in persons with asthma.

Dobutamine and **xameterol** are selective β_1-adrenoceptor agonists, whose inotropic properties are used to increase cardiac output in heart failure and other forms of shock.

Isoprenaline is a less selective β-adrenoceptor agonist with both chronotropic (increasing heart rate) and inotropic (increasing force of cardiac contraction) actions. It can be used as a temporary measure in states of bradycardia and heart block, before a cardiac pacemaker is fitted.

(c) Toxic effects of catecholamines

Cardiac arrhythmias which can lead to cardiac arrest are a major toxic effect of β-adrenoceptor stimulants when these substances gain access to the heart. Any parenteral administration of such compounds carries this risk. Substances which do not have the same potential for initiating cardiac arrhythmias are a logical alternative – for example, a non-catecholamine vasoconstrictor or a selective β_2-adrenoceptor agonist. Certain volatile general anaesthetics can sensitize the heart to sympathomimetic amines; thus, following cyclopropane, chloroform, halothane or trichlorethylene, the use of directly or indirectly-acting sympathomimetics should be avoided, as the risk of cardiac arrhythmias occurring is increased.

(d) α-Adrenoceptor blockers

Prazosin is used in the treatment of hypertension but it can cause an unacceptable degree of orthostatic hypotension. **Phentolamine** and **phenoxybenzamine** are used in the diagnosis and surgical removal of phaeochromocytoma, in which situation they prevent an excessive rise in blood pressure. Whereas **phentolamine** has a short duration of action and must be given intravenously, **phenoxybenzamine** has a prolonged action and can be given orally or intravenously.

Labetolol is an antihypertensive agent which has both α- and β-adrenoceptor blocking properties.

Chlorpromazine and other antipsychotics which have some α-adrenoceptor blocking properties can cause hypotensive side effects, but they are not used for this purpose.

The side effects of α-adrenoceptor blockers include nasal stuffiness, sedation, tachycardia, orthostatic hypotension and dizziness.

(e) β-Adrenoceptor blockers

β-Adrenoceptor blockers are used in the treatment of hypertension, cardiac arrhythmias, tremor in thyrotoxicosis, prophylaxis of migraine and angina. They include (−)-**propranolol, atenolol, oxprenolol** and **sotalol**. In all these conditions they reduce the effects of catecholamines which are expressed by activation of β-adrenoceptors. Mechanisms of the antihypertensive actions of β-adrenoceptor blockers are discussed in section 3.4.5. β-Adrenoceptor blockers do not cause postural hypotension.

β-Adrenoceptor blockers can restore normal electrical activity in the heart, when this has been disrupted by excess sympathetic activity, by treatment with cardiac glycosides, or by poor oxygenation. It is not easily possible to rationalize the effectiveness of β-adrenoceptor blockers in restoring normal heart rhythm in response to such diverse noxious stimuli. It appears that both β-adrenoceptor blocking action and local anaesthetic effects contribute to the overall efficacy of these drugs in this condition.

β-Blockers are used to treat dysrhythmias arising from excess stimuli from the sino-auricular and sino-ventricular nodes. Recent studies have shown that long term use of β-blockers (up to 1 year), reduces mortality and morbidity following myocardial infarction.

Angina is treated with β-adrenoceptor blockers because they reduce or prevent ischaemic pain caused by sympathetic stimulation of the heart which arises in response to emotional or physical stress. This increases the oxygen demand of the heart, which cannot be met by the defective coronary circulation, and pain follows. Treatment with β-adrenoceptor blockers depends on their ability to block sympathetic drive, and is independent of membrane-stabilizing action. Treatment is given prophylactically, and other agents (e.g. nitrates) can be used to abort individual attacks.

β-Adrenoceptor blockers are used to control many forms of tremor. Tremor and tachycardia which occur in thyrotoxicosis (a state in which sympathetic function is potentiated), in phaeochromocytoma, and in some anxiety states, can also be controlled by β-adrenoceptor blockers. Whether there is a CNS component to the anxiolytic action of these compounds is not established; there are indications that they can cause depression in some patients.

(f) Adverse effects of β-adrenoceptor blockers

In individuals with limited 'cardiac reserve', i.e. those who cannot increase cardiac output in response to work, or in those who rely on a degree of sympathetic tone to maintain adequate cardiac output, β-adrenoceptor blockers can induce heart failure by further reducing the ability of the heart to respond to sympathetic stimulation. As β-adrenoceptor blockers are not selective for β_1- or β_2-adrenoceptors, they can precipitate bronchoconstriction in persons with asthma, as well as preventing bronchodilation by β_2-adrenoceptor agonists. Patients complain of cold extremities (hands and feet); this is probably due to loss of β-adrenoceptor-mediated vasodilation. β-blockers may need to be used with caution in persons suffering from diabetes because activation of β-adrenoceptors is necessary for mobilization of glucose during stress, and should not be given to patients with peripheral vascular disease.

3.6.5 Inactivation of released noradrenaline

Drugs blocking neuronal noradrenaline uptake

Indirectly-acting sympathomimetic amines, cocaine, certain MAO inhibitors and the tricyclic antidepressants inhibit noradrenaline re-uptake. Only the latter two classes are used in the treatment of depression and, whereas they do not have overt stimulation actions, they elevate the mood of the depressed patient back within the normal range. This is a fundamental difference between clinically used antidepressants and **amphetamine, cocaine** and **methylxanthine** stimulants.

Depression is characterized by a feeling of sadness, and it frequently occurs following bereavement. In such circumstances it is the normal response to a well-identified situation. If depression occurs as a response to an identifiable cause, it is called exogenous or reactive depression, and this type of depression does not respond well to drug therapy, but does respond to psychotherapy, counselling and changes in environment. Endogenous depression does not have a readily identifiable cause, and can start without any obvious precipitating factor. This latter type of depression does not respond to psychotherapy or counselling, but is responsive to drug therapy. Electroconvulsive therapy (ECT) is most effective when used to treat endogenous depression; indeed this is the only indication for the use of this empirical treatment. Tricyclic antidepressants and MAO inhibitors must be administered for 1–3 weeks before a beneficial effect is seen.

3.6.6 The biogenic amine theory of depression

The mechanism of action of tricyclic antidepressants, MAO inhibitors, and other agents used to treat depression is far from clear. The biogenic amine theory of depression was developed over 30 years ago from laboratory and clinical observations with **reserpine**, MAO inhibitors and tricyclic compounds. **Reserpine** caused sedation in animals, and a 'depressive' syndrome in man; it also caused a decrease in the amount of noradrenaline and 5-hydroxytryptamine (and dopamine) in the central (and peripheral) nervous system in both man and animals. So depression became associated with low levels (concentrations) of these chemicals in the brain, and this was extrapolated to mean that there was underactivity of these neurochemical systems. MAO inhibitors (section 3.5.3) were found to prevent reserpine-induced sedation in animals, were found to reverse depression in some clinical trials, and in biochemical studies, they were found to increase the concentration of noradrenaline, 5-hydroxytryptamine (and dopamine) in nervous tissues. Thus reversal of depression (or an anti-depressant effect) became associated with an increase in the concentration of monoamines in the brain, and this was extrapolated to mean that an apparent increase in the activity of these neurochemical systems could bring about a reversal of depression.

Some years later, tricyclic antidepressants (as they are now known) were found to block the re-uptake of noradrenaline and 5-hydroxytryptamine into neurones. In controlled clinical trials, these compounds were found to have a true antidepressant action, and these findings were used in support of the theory which had arisen on the basis of observations with reserpine and MAO inhibitors. The tricyclic compounds blocked neuronal re-uptake, and this became associated with the idea that if more amine was available in the synapse for a longer time than usual (because the major inactivation process was stopped), then this resulted in extra stimulation of receptors, and this brought about a reversal of depression. Stated more concisely, the biogenic amine theory of depression states that in endogenous depression, there is a deficit of stimulation of some receptors in the CNS, and antidepressants return the level of receptor stimulation to that which is closer to the normal (non-depressed) level.

The theory has been criticized for its relatively simplistic nature, and for the selectivity of observations upon which it depends. MAO inhibitors frequently have actions other than just inhibiting MAO, and tricyclic antidepressants block many receptors (at concentrations far lower than those needed to inhibit amine uptake) in addition to blocking amine uptake. A further disparity between theory and observation is the time scale of events following drug treatment. The biochemical effects can be measured within hours of starting drug treatment; antidepressant effect is only seen after some 2–3 weeks of drug administration.

More recent work has examined biochemical events other than inhibition of MAO and amine uptake. Drug-binding studies have shown that the number of β-blocker binding sites in the CNS decreases following 2–3 weeks of treatment with antidepressants or following ECT (electroconvulsive therapy). Other ligand binding studies have shown that some sites selectively labelled by spiroperidol (which has been proposed to be a dopamine and 5-HT binding site ligand) also decrease in number in the CNS following chronic treatment with a wide range of antidepressants.

It is currently suggested that there is a change in either sensitivity or numbers of certain receptors in depression, and that antidepressants return these changes towards 'normality'. The exact anatomic location of these binding sites, and their relationship to any neurotransmitter receptors still has to be established.

Thus the neurotransmitter receptor hypothesis of antidepressant drug action proposes that depression and its reversal is mediated through changes in adrenergic (and 5-HT) binding sites.

Tricyclic antidepressants have actions other than the relief of depression, both when given alone and when given with other drugs. The side effects of tricyclic antidepressants are usually undesirable; however, some may be sufficiently important that they determine choice of compound. Thus, whereas **imipramine** has only weak sedative actions, **amitriptyline** has marked sedative effects, and so the latter may be more suitable for the treatment of agitated patients. **Desipramine** and **protriptyline** have mild stimulant actions, and may therefore be useful in persons who appear in need of arousal. Tricyclic antidepressants have been used in the treatment of bed wetting (enuresis) in children. The effect only lasts while the drug is given, and the use of the drugs in this condition does not seem well-founded. Tricyclic antidepressants are used with opioid analgesics in the control of pain. Whereas the mechanism is not understood, useful potentiation of the analgesia is obtained.

Imipramine and **amitriptyline** are metabolized *in vivo* to their respective demethylated derivatives, **desipramine** and **nortriptyline**. All four compounds are used as antidepressants, but there does not appear to be any advantage (such as speed of onset of antidepressant effect) in the demethylated derivatives. The common incidence of adverse side effects to MAO inhibitors and tricyclic antidepressants (see below), has led to a search for effective antidepressants which do not have such adverse effects.

Mianserin, **viloxazine** and **trazodone** are non-MAO inhibitors, non-tricyclic antidepressants whose mechanism(s) of action are not clearly established. They do not give rise to adverse effects which occur with MAO inhibitors, and they have a lower frequency of anticholinergic and cardiotoxic side effects. The actions of **mianserin** can be reversed by **clonidine** (as they both act at presynaptic α_2-receptors).

Low doses of some antipsychotic compounds (sections 4.4.4 and 4.6.4) such as **fluphenazine** have been used as adjuncts to treatment with tricyclic antidepressants. They may be effective by reducing agitation (which often occurs together with depression), but whether they have an antidepressant action is not established.

Benzodiazepines (section 7.1.4) are sometimes used in 'mild depression', possibly because of their mild mood-elevating actions, but again it is not clear whether they have a true antidepressant action, or their effectiveness is due to a reduction of anxiety. The role of 5-HT in depressive illness is discussed in section 5.6.4.

(a) Side effects of tricyclic antidepressants

Most tricyclic antidepressants can block muscarinic receptors (anticholinergic action); thus they frequently cause memory disorders, blurring of vision, dry mouth, difficulty in micturition and constipation. They can also weakly block α-adrenoceptors, which may cause postural hypotension. While it may seem paradoxical, hypertension can also be caused by tricyclic compounds, presumably because they prolong α- and β-adrenoceptor stimulation following inhibition of noradrenaline re-uptake, and this increases vasoconstriction and cardiac output. Tachycardia, palpitations, development of a fine tremor and weight gain are other side effects.

Within the CNS, the side effects of tricyclic antidepressants are complex and not well understood. In patients with epilepsy, they can precipitate seizures, and this has been known to occur in persons without a previous history of epilepsy. The tranquillizing action has been mentioned above. With regard to their interactions with barbiturates, hypnotics and ethanol, they must be regarded as general CNS depressants, since they can cause serious respiratory depression if taken in overdose or in combination with the above drugs.

As tricyclic antidepressants inhibit the major mechanism for terminating the activity of released noradrenaline, the sympathomimetic actions of catecholamines can be dangerously potentiated (section 3.6.4). Tricyclic antidepressants have a higher affinity for the plasma membrane of noradrenergic neurones than do the adrenergic neurone blockers; thus, the effectiveness of the latter is decreased by tricyclic antidepressants (section 3.6.3).

Concurrent use of tricyclic antidepressants with MAO inhibitors has been reported to have caused death following hyperpyrexia, coma and convulsions. While such drug combinations are recommended by some authorities, the seriousness of the reaction when it occurs would suggest that the two classes of antidepressants should not be used together.

(b) MAO inhibitors

MAO inhibitors are not as effective as tricyclic antidepressants in relieving endogenous depression, but they have been found of value in some forms of exogenous depression. Numerous restrictions are imposed on the patients with regard to avoidance of certain foods and medicines, and this makes the drugs dangerous in use, especially in persons who may not be in full control of all their actions.

Tranylcypromine and **phenelzine** have marked sympathomimetic actions, which can cause hypertension, tremor and tachycardia. In addition to inhibiting MAO, they also inhibit noradrenaline uptake, and thus have some actions in common with the tricyclic antidepressants. **Tranylcypromine** releases dopamine as well as blocking its uptake; it thus has some **amphetamine**-like actions. This effect probably accounts for the relatively rapid onset of action of this compound when compared with MAO inhibitors without sympathomimetic actions such as **iproniazid** and **isocarboxazid**. Iproniazid can cause dangerous liver toxicity.

(c) Interactions of MAO inhibitors

As more noradrenaline is stored in nerve terminals during inhibition of MAO, the effect of indirectly acting sympathomimetic amines is potentiated by these compounds. **Amphetamines** and any nasal decongestants or bronchodilators which contain indirectly-acting sympathomimetic amines must not be used by persons taking MAO inhibitors, as hypertensive responses will occur.

Tyramine is an indirectly-acting sympathomimetic amine found in many foods, but especially those which are prepared by the actions of micro-organisms (e.g. cheeses, alcoholic drinks, yeast extracts). Normally, **tyramine** is destroyed in the intestine and liver by MAO, but when MAO is inhibited (and MAO inhibitors inhibit MAO at all sites in the body), **tyramine** passes into the bloodstream. On reaching sympathetic nerve endings, **tyramine** causes noradrenaline release, which leads to activation of both α- and β-adrenoceptors. This causes vasoconstriction and increases cardiac output, leading to hypertension and possibly a hypertensive crisis. This process is sometimes referred to as the 'cheese reaction'. Stroke, subdural and subarachnoid haemorrhage and headache might occur during such a hypertensive crisis. Treatment

Table 3.1 Summary of drugs which modify noradrenergic transmission

Mechanism	Drug	Effect	Uses
Synthesis	α-Methyl-DOPA α-Methyl-*p*-tyrosine Disulfiram	Synthesis of false transmitter Inhibits tyrosine hydroxylase Inhibits dopamine-β-hydroxylase	Antihypertensive Experimental Alcoholism
Storage	Reserpine Tetrabenazine	Disrupt NA storage	Antihypertensive
	MAO inhibitors	Enhance NA storage	Antidepressants
Release	(+)-Amphetamine Tyramine	Indirectly-acting sympathomimetic amines, cause release of NA	Elevation of blood pressure Interactions with MAO inhibitors
	Guanethidine Debrisoquine Bethanidine Bretylium	Adrenergic neurone blockers, decrease NA release	Antihypertensives Cardiac arrhythmias
Receptors	NA Phenylephrine	α_1 selective agonists	Nasal decongestant
	α-Methyl-NA Clonidine	α_2 selective agonists	Hypotensives on central α-adrenoceptors
	Phentolamine Phenoxybenzamine	Non-selective α blockers	Antihypertensives

Drug	Classification	Clinical use
Prazosin	α_1 selective blockers	
Yohimbine	α_2 selective blockers	
NA **Adrenaline** **Isoprenaline**	Non-selective β-agonists	
Salbutamol **Terbutaline**	β_2 selective agonists	Bronchodilators
Dobutamine	Selective β_1 agonist	
(−)-Propranolol **Oxprenolol**	Non-selective β-blockers	Antihypertensives Anti-anginal Antidysrhythmics
Sotalol **Alprenolol**	β_1 selective blockers	
(+)Amphetamine **Cocaine**	Neuronal-uptake inhibitors	Stimulants
Imipramine **Amitriptyline**		Antidepressants
Iproniazid **Nialamide** **Phenelzine** **Tranylcypromine**	MAO inhibitors	Antidepressants

Inactivation of uptake

of metabolism

of such hypertensive crises is most rationally done by symptomatic use of adrenoceptor blockers.

MAO inhibitors must not be given with **L-DOPA,** as hypertension and cardiac arrhythmias can occur, presumably because of the increased availability of noradrenaline. Broad beans, which are a good source of L-DOPA, must not be eaten by persons taking MAO inhibitors.

MAO inhibitors also inhibit other enzymes, in particular the microsomal enzymes concerned with the metabolism of many drugs. Thus the effects of **pethidine, barbiturates** and other CNS depressants may be prolonged and potentiated in persons taking MAO inhibitors. The potentiation of the respiratory-depressant actions of pethidine and barbiturates can prove fatal. As MAO inhibitors have a long duration of action, it is usually not safe to use drugs which interact with MAO inhibitors until two weeks after the last dose of MAO inhibitor. MAO inhibitors may also potentiate the actions of insulin and some oral hypoglycaemic agents, presumably by inhibiting their metabolism by enzymes. The simultaneous use of MAO inhibitors and tricyclic antidepressants has been discussed above (p. 82).

A summary of the drugs which modify noradrenergic transmission is given in Table 3.1.

FURTHER READING

Dahlstrom, A., Belmaker, R.H. and Sandler, M. (1988) *Progress in Catecholamine Research. Part A: Basic Aspects and Peripheral Mechanisms. Part B: Central Aspects. Part C: Clinical Aspects.* Alan R. Liss, New York.

Gilman, A.G., Rall, T.W., Nies, A.S. and Taylor, P. (1990) *Goodman & Gilman's The Pharmacological Basis of Therapeutics,* Chapters 10 and 11. Pergamon Press, New York.

Heal, D.J. and Marsden, C.A. (1990) *The Pharmacology of Noradrenaline in the Central Nervous System.* Oxford University Press, Oxford.

Sibley, D.R., Benovic, J.L., Caron, M.G. and Lefkowitz, R.J. (1987) Regulation of transmembrane signalling by receptor phosphorylation. *Cell,* **48,** 913–22.

4 Dopamine

Dopamine is a catecholamine neurotransmitter in the central nervous system (CNS) and at some ganglia in the autonomic nervous system; most of our knowledge of the functional role of dopamine relates to the CNS. In both the peripheral and central nervous system, dopamine is a precursor of noradrenaline and adrenaline.

4.1 SYNTHESIS

Dopamine is made from the amino acid L-tyrosine, which is hydroxylated by the enzyme tyrosine hydroxylase to 3,4-dihydroxyphenylalanine (L-DOPA). Tyrosine hydroxylase is found in the cytoplasm of catecholamine neurones and requires reduced pteridine cofactor, ferrous ions and oxygen. Tyrosine is taken up into dopaminergic neurones from plasma by an active-transport process found on the nerve membrane. L-DOPA is decarboxylated by L-aromatic amino acid decarboxylase (DOPA decarboxylase) to form dopamine (Fig. 4.1). DOPA decarboxylase is a cytoplasmic enzyme which requires pyridoxal phosphate (vitamin B_6) as cofactor.

4.1.1 Control of dopamine synthesis

The rate-limiting step in the synthesis of dopamine is the conversion of tyrosine into L-DOPA by tyrosine hydroxylase. Under normal conditions, this enzyme is completely saturated with L-tyrosine, and thus increases in circulating tyrosine levels do not increase the rate of dopamine synthesis. High intraneuronal concentrations of dopamine inhibit tyrosine hydroxylase by end-product inhibition, thus decreasing the rate of dopamine synthesis. The description of the regulation of the activity of tyrosine hydroxylase by phosphorylation in section 3.1.1 applies equally to dopamine.

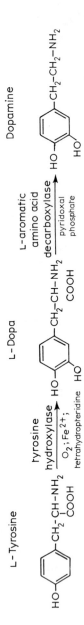

Figure 4.1 Synthesis of dopamine.

4.1.2 Increase of dopamine synthesis. L-DOPA

L-DOPA is actively taken up into dopamine neurones in the CNS, where it is converted into dopamine by DOPA decarboxylase. There is a significant increase in the amount of dopamine stored and available for release following L-DOPA therapy. L-DOPA can also be taken up by glial cells where it is decarboxylated by DOPA decarboxylase. Glial cells cannot store dopamine efficiently, so that the dopamine leaks out and is available for the stimulation of dopamine receptors and for uptake into dopamine neurones. Decarboxylation of L-DOPA at non-neuronal sites is believed to be important in conditions where there is degeneration of dopamine neurones (see the discussion of Parkinson's disease in section 4.6.1). By comparison with the dopaminergic system there is relatively little increase in the synthesis of noradrenaline following L-DOPA administration.

4.1.3 Inhibition of dopamine synthesis

α-Methyl-*p*-tyrosine competitively inhibits tyrosine hydroxylase, and therefore inhibits the synthesis of all catecholamines – including dopamine. It is used experimentally. Peripherally-acting inhibitors of DOPA decarboxylase, such as **carbidopa** and **benserazide**, are important when only the central conversion of L-DOPA into dopamine is required, as in the drug therapy of Parkinson's disease.

Carbidopa (α-methyl-dopa-hydrazine) is not sufficiently lipid soluble to cross the blood–brain barrier. NSD 1015 is a DOPA-decarboxylase inhibitor which is lipid soluble and inhibits DOPA-decarboxylase both peripherally and in the CNS; it is of experimental interest only.

4.2 STORAGE

The storage of dopamine has many features in common with the storage of noradrenaline. Thus, dopamine is found in storage granules in which it is complexed with chromogranins, divalent metal ions and ATP. A small fraction of the total stored dopamine appears not to be complexed in storage granules. What has been stated about the possible functions of the various proposed storage forms of noradrenaline (section 3.2) probably also applies to dopamine.

4.2.1 Disruption of dopamine storage

By mechanisms similar to those described for noradrenaline (section 3.2.2), the uptake of dopamine into storage vesicles and the stability of the storage complex is disrupted by **reserpine** and **tetrabenazine**. As a result less dopamine is stored in nerve terminals, and less is available for release. The disruption of dopamine storage in man may result in a Parkinsonian syndrome and depression. **Tetrabenazine** has actions similar to those of **reserpine**, but a shorter duration of action.

4.3 RELEASE

Dopamine is released into the synaptic cleft by exocytosis; this is a calcium dependent process and occurs in response to depolarization of the axon terminal.

Recently synthesized dopamine is released in response to action potentials; release from this store is inhibited by pre-treatment with α-methyl-p-tyrosine (an inhibitor of the rate limiting enzyme tyrosine hydroxylase). Dopamine which has been synthesized but which was not released is transferred to a reserve store. The reserve store is not available for release under normal circumstances, but some drugs appear to be able to transfer the dopamine from the reserve store to the releasable store. Both the releasable and the reserve store are susceptible to disruption by **reserpine**. The two dopamine stores cannot be seen by anatomical studies; their existence is inferred from pharmacological experiments with indirectly acting sympathomimetic amines, **reserpine**-like drugs and α-methyl-p-tyrosine.

Amphetamine-like drugs can displace dopamine from the storage vesicles into the cytoplasm. The dopamine is not destroyed in the cytoplasm because **amphetamine** is a reversible inhibitor of monoamine oxidase. Dopamine diffuses into the extracellular space down a concentration gradient; alternatively it may enter the synapse by an exchange diffusion process. Displacement of dopamine by **amphetamine** and its subsequent release is not a calcium dependent process.

Dopamine release can be modulated by pre-synaptic axon terminal dopamine D2 receptors. These receptors are linked by an inhibitory G protein to adenylyl cyclase; activating these receptors leads to decreased synthesis of cyclic AMP. In a manner analogous to that which has been described for the α_2-adrenoreceptor, it is suggested that decreased phosphorylation of calcium/calmodulin sensitive protein kinase 2 leads to decreased coupling of dopamine vesicles to the sites of exocytosis thus decreasing dopamine release (section 3.3.1).

5-HT3 receptors on dopamine axon terminals may act as heteroreceptors which facilitate dopamine release (section 5.5).

4.4 DOPAMINE SYSTEMS

Dopaminergic neurotransmission has been shown to occur only in the CNS. Many peripheral tissues (e.g. gut, blood vessels, heart) respond to exogenously-applied dopamine, but dopaminergic innervation of these tissues has not been found. Experiments have shown that substances which act as agonists or antagonists on the dopamine systems in the CNS can act in what appears to be a similar selective fashion on peripheral tissues. This has led to the conclusion that there may be dopamine receptors in peripheral tissues which have a dopaminergic innervation which is as yet poorly described.

4.4.1 Dopamine receptors at sites outside the CNS

In the peripheral nervous system dopamine receptors are believed to be located in some sympathetic ganglia, exocrine glands and the gastro-intestinal tract and on mesenteric and renal arteries. In the sympathetic ganglia of the autonomic nervous system, dopamine receptors linked to an adenylate cyclase appear to cause changes in the permeability of the postganglionic neuronal membrane, leading to inhibitory postsynaptic potentials which are slow in onset. This leads to inhibition of transmission at these ganglia.

In renal and mesenteric arteries, dopamine causes a dose-dependent vasodilator response, which can be attenuated by pretreatment with dopamine-receptor blocking agents. These responses to dopamine are not affected by α- or β-adrenoceptor blockers. Relaxation in response to dopamine has been reported to occur in isolated preparations of several regions of the gastro-intestinal tract, and increased secretion of enzymes from the salivary glands and the pancreas has been demonstrated in response to dopamine agonists. The physiological significance of dopamine at peripheral sites has not been established.

Dopamine is also found in the glomus cells of the carotid body, where it appears to have a role in control of respiratory reflexes. Hypoxia decreases dopamine release in the carotid body; this removes the hyperpolarization of sensory neurones and reflexly stimulates respiration.

4.4.2 Dopamine and its receptors in the CNS

Three major neuronal systems in the brain which use dopamine as a neurotransmitter have been discovered (see Fig. 4.2). The identification of areas rich in dopaminergic nerve terminals has led to the association of those areas with functions which may be under the control of dopaminergic neurones.

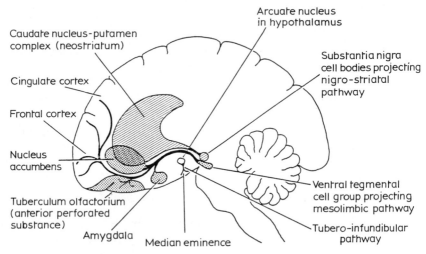

Figure 4.2 Dopamine pathways in the CNS.

(a) Dopaminergic nigro-neostriatal pathway

The nerve-cell bodies lie in the substantia nigra (midbrain) and send axons which terminate in the caudate nucleus–putamen complex (the neostriatum). This system and these brain areas are concerned with integration of incoming sensory stimuli and control of movement. It is part of the extrapyramidal system of the basal ganglia.

(b) Dopaminergic midbrain mesolimbic forebrain system

The nerve-cell bodies lie in the ventral tegmentum of the midbrain, medial to and within the substantia nigra. The axons from the cell bodies terminate in the head of the caudate nucleus, the nucleus accumbens, the tuberculum olfactorium (corresponding possibly to the anterior perforated substance of human brain), the amygdaloid nuclei, frontal and cingulate cortex, and parts of the neocortex. This system and the associated brain areas are believed to be concerned with cognitive, reward and emotional behaviour.

(c) Dopaminergic tubero-infundibular system

The nerve-cell bodies lie in the region of the arcuate nucleus of the hypothalamus, with short axons which terminate in the median eminence. This system and the associated brain area are concerned with neuronal control of the hypothalamic–pituitary endocrine system.

Dopamine and dopamine receptors linked to an adenylate cyclase system are present in the retina. The functional role of dopamine at this site is not fully established; a role in regulation of spread of dendrites is indicated. Depletion of dopamine in the retina is associated with photophobia or exaggerated responses to illumination.

4.4.3 Dopamine receptors

Three classes of dopamine receptor have been identified on the basis of pharmacological, transducer and effector systems, and gene sequences.

The dopamine D1 receptor (which appears to be indistinguishable from the DA1 receptor found peripherally) belongs to the G protein receptor super family and agonists acting at this receptor stimulate the production of cyclic AMP. SKF38393 and **fenoldopam** are selective agonists at D1 receptors; SCH23390 is a selective competitive antagonist at D1 receptors.

One of the functions of cyclic AMP generated as a result of D1 receptor stimulation is the phosphorylation of a protein called DARPP-32 (dopamine and cyclic AMP regulated phosphoprotein molecular weight 32 000). Phosphorylated DARPP-32 is an inhibitor of protein phosphatase 1, an enzyme which dephosphorylates substrate proteins which include structural proteins, ion channel proteins and proteins associated with synaptic vesicles. Regulation of protein phosphatase 1 by D1 receptors and other secondary messengers (section 8.4.1: NMDA receptors) can affect a wide range of effector systems in cells with D1 receptors.

Gene sequence studies for D2 receptors have revealed that there may be two forms of the D2 receptor. The two forms of the D2 receptor differ in the number of the amino acids found in the third cytoplasmic domain of the receptor protein which has either 415 or 441 amino acids. The third cytoplasmic domain of these receptors appears to be the site where G protein is bound. It is not at present known whether different G proteins bind to the two sub-types of D2 receptor protein. Two transducer effector systems are associated with the D2 receptor. The first described involved a Gi protein which inhibits adenylyl cyclase and leads to decreased cyclic AMP production. The second is linked by a G protein to a potassium ion channel.

Quinpirole is a selective agonist at D2 receptors; $(-)$ **sulpiride** is a selective D2 receptor blocker.

All D1 receptors are located on postsynaptic structures; D2 receptors are found both pre- and post-synaptically, and all dopamine auto-receptors whether on somatodendritic sites or on axon terminals are of the D2 type. Pharmacological studies indicate that there might be regional variations in the distribution of D2 receptor sub-types. A major research effort is currently directed towards the discovery of selective and specific D1 and D2 receptor agonists and blockers.

Dopamine is a D1 and D2 agonist, and behavioural studies indicate that activation of both D1 and D2 is essential for expression of many dopamine functions in the CNS. The pharmacology of D1 and D2 receptor mediated behaviours and functions in the CNS is incompletely described. Studies in animals with selective D1 and D2 antagonists are not yet complete, and experience of such drugs in man is very limited. The results of such experiments are eagerly awaited as they could add valuable insights into mechanisms to consciousness, cognition, thought, emotional behaviour and memory.

A dopamine D3 receptor has also been described; selective ligands for the D3 receptor are not available (yet), but the D3 receptor is found mainly in the mesolimbic pathways associated with cognitive and emotional functions. It is likely that antipsychotic drugs interact at this site as well as at D2 receptors.

In addition to **dopamine**, there are a number of synthetic dopamine receptor agonists including **epinine** (N-methyldopamine), **apomorphine** and its analogue **N-propylnoraporphine, piribedil** (ET495), certain ergot alkaloids including **ergometrine, bromocriptine** (2-bromo-α-ergo-cryptine), **lysuride, pergolide** and **lysergic acid diethylamide** (LSD), and the pro-drug **ADTN** (2-amino-6,7-dihydroxy-1,2,3,4-tetrahydro-naphthalene). Many of these agents are used in, or have been assessed for, the treatment of Parkinson's disease (section 4.6.4). None of these drugs exhibits total selectivity for either D1 or D2 receptors; they will all, for example, to a certain extent, stimulate adenylate cyclase, although the ergot derivatives are more active at D2 sites.

Indirectly-acting dopamine agonists – for example **L-DOPA** and (+)-**amphetamine** and its derivatives – have been considered above (sections 4.1.2 and 4.3).

4.4.4 Dopamine receptor blockers

The chemical classes of dopamine receptor blockers are summarized in Table 4.1.

Dopamine receptor blockers which are effective in the treatment of schizophrenia are referred to as antipsychotics. They are also referred to as neuroleptics or major tranquillizers. Dopamine receptor blockers have anti-emetic activity, which means they can reduce nausea and vomiting caused by chemicals or radiation. Motion sickness and vertigo are usually unresponsive to dopamine receptor blockers. Dopamine receptor blockers have a common property of causing sedation without inducing

Table 4.1 Dopamine receptor blockers

Chemical class	Drug name	Comment
Phenothiazines	Chlorpromazine Thioridazine Fluphenazine	No selectivity for D1 or D2 receptors
Butyrophenones	Haloperidol Spiroperidol	Some selectivity for D2 receptors
Thioxanthenes	Chlorprothixene Flupenthixol	Little selectivity for D1 or D2 receptors
Diphenylbutylpiperidines	Pimozide Penfluridol	Some selectivity for D2 receptors
Dibenzodiazepines	Clozapine	Some selectivity for D2 receptors
Substituted benzamides	Metoclopramide Domperidone Sulpiride	Selective for D2 receptors
	SCH 23390	Selective for D1 receptors

readily reversible loss of consciousness – 'sleep'; they do not possess hypnotic activity. This sedative action is useful in the treatment of severe agitation or mania.

4.5 INACTIVATION OF DOPAMINE

4.5.1 Neuronal uptake of dopamine

Following release into the synaptic cleft, the biological activity of dopamine is terminated by a neuronal re-uptake system which is similar in many ways to the uptake 1 system described for noradrenaline (section 3.5.1). This is a high-affinity, energy-dependent active-transport system.

Inhibitors of neuronal uptake of dopamine

Among the drugs which inhibit neuronal dopamine uptake are the indirectly-acting dopamine-receptor agonists, for example, (+)-**amphetamine**; certain antimuscarinic drugs, for example, **benztropine** and **benzhexol**; and other compounds including **nomifensine, tyramine** and **tranylcypromine**. In general, the latter substances are more potent inhibitors of dopamine re-uptake than they are releasers of dopamine. The biological activity of these compounds at dopaminergic synapses probably depends on both uptake inhibition and release of dopamine.

Tricyclic antidepressants are weak inhibitors of dopamine uptake; they have some selectivity for noradrenaline or 5-HT uptake. **Cocaine** blocks

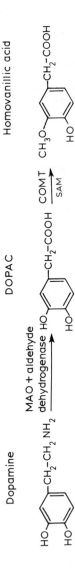

Figure 4.3 Metabolism of dopamine. SAM, 5-adenosyl-L-methionine.

uptake of both dopamine and noradrenaline; GBR 12909 is a highly selective dopamine uptake blocker which is used experimentally.

4.5.2 Enzymatic destruction of dopamine. MAO and COMT uptake

Any dopamine that is taken up into the neurone, and which is not bound into the storage granules, will be metabolized by monoamine oxidase (MAO) and by catechol-O-methyl transferase (COMT). MAO is located mostly on mitochondria, while COMT is a soluble enzyme found in the cytoplasm (section 3.5.2). The major metabolites of dopamine (Fig. 4.3) are 3,4-dihydroxyphenylacetic acid (DOPAC) and homovanillic acid (HVA, 3-methoxy-4-hydroxyphenylacetic acid).

(a) Inhibitors of dopamine degradation by MAO

MAO inhibitors (**iproniazid, tranylcypromine, nialamide, phenelzine**) act irreversibly and are used as antidepressants. The amount of dopamine found in nerve terminals is increased following treatment with a MAO inhibitor, and more is available for release. **Selegiline** is only a selective inhibitor of MAO B in low doses. If higher doses are given then it also inhibits MAO A. At the low doses needed to selectively inhibit MAO B, the use of **selegiline** does not impose a dietary restriction (avoidance of foods containing **tyramine**) which is inherent in the use of non-selective MAO A inhibitors (section 3.6.5).

(b) MPTP

When man or other primates take the neurotoxin MPTP they develop the symptoms of Parkinsons' disease. MPTP is converted via intermediates to MPP^+ by MAO B, and MPP^+ is taken up and concentrated in mitochondria where it inhibits the respiratory chain enzymes and eventually causes death of the cell. The selective neurotoxicity of MPTP for dopaminergic neurones may arise from the selective location of MAO B in dopamine neurones. If the animals are pre-treated with selegiline prior to MPTP then they are protected from the onset of the Parkinsonism syndrome.

4.6 THERAPEUTIC APPLICATIONS AND CONSEQUENCES OF DRUGS ACTING AT DOPAMINERGIC SYNAPSES

4.6.1 Drugs which affect dopamine synthesis

(a) L-DOPA and Parkinsons's disease

L-DOPA is taken up by dopaminergic neurones, where it is made into dopamine, and this is the basis for the use of **L-DOPA** in the treatment of

Parkinson's disease. Parkinson's disease is a neurological disorder characterized by rigidity of limbs, trunk and face, a tremor when awake and resting, abnormal body posture and an inability to initiate voluntary motor activity (akinesia). It occurs mainly in elderly people, and may be of unknown cause, or a result of cerebral injury or disease. The condition is associated with a degeneration of dopamine neurones, predominantly in the nigro-neostriatal pathway, and decrease in the amount of dopamine and melanin in the substantia nigra and neostriatum. Parkinson's disease is a chronic and progressive degenerative disorder and, as with many neurological states, drug therapy cannot cure the disease; it only alleviates the clinical symptoms. A Parkinsonian syndrome (drug-induced Parkinsonism) can also occur as a side effect of treatment with **reserpine** or neuroleptic drugs (sections 4.6.2 and 4.6.4).

More recently a newer, often younger group of individuals has been recognized who exhibit Parkinsonian-like symptoms following exposure to a neurotoxic agent MPTP. This is discussed more fully in section 4.6.5.

The association of decreased dopaminergic function and Parkinson's disease was first made when patients treated with **reserpine** (for hypertension) developed a Parkinsonian syndrome. It was known from animal experiments that **reserpine** caused a great decrease in the concentrations of dopamine, noradrenaline and 5-hydroxytryptamine in the brain, and that the animals became akinetic (did not walk). This akinesia was reversed if the animals were treated with **L-DOPA**, and brain dopamine concentrations returned towards normal values. When **L-DOPA** was tested in people with Parkinson's disease it was found to be particularly effective in restoring the ability to initiate movement, and it is currently the most effective treatment for this condition. Dopamine itself cannot be given because it is not sufficiently lipid soluble to enter the brain, and because it is rapidly destroyed by MAO in the intestine and liver.

L-DOPA is given by mouth, is absorbed from the gastrointestinal tract and is carried to the brain by the bloodstream. **L-DOPA** is taken up not only by dopaminergic nerve terminals, but also by some non-neuronal tissues such as glial cells. Intracellularly, **L-DOPA** is decarboxylated to dopamine and, while it can be stored effectively within remaining dopaminergic neurones, this does not occur in glial cells; from these it appears that dopamine may diffuse into the synaptic cleft and act on dopamine receptors. Only a small fraction of the total oral dose of **L-DOPA** reaches the brain, where it can exert its beneficial action. Much is taken up into the liver and other non-neuronal tissues outside the CNS, where it is decarboxylated to dopamine; this is then inactivated by MAO and COMT. Because of this, high doses of **L-DOPA** may be needed in order that adequate therapeutic response occurs. Unfortunately, as the dose of **L-DOPA** is increased, so the incidence of side effects increases – on occasions to the extent that **L-DOPA** therapy has to be discontinued.

It is possible to decrease the dose of **L-DOPA** needed to control Parkinson's disease – and consequently to decrease the incidence of side effects – by preventing the decarboxylation of **L-DOPA** at sites outside the CNS. This can be achieved by concurrently administering inhibitors of DOPA decarboxylase which do not cross the blood–brain barrier. Such peripheral DOPA decarboxylase inhibitors include **carbidopa** and **benserazide**, and they are given together with **L-DOPA**. Less **L-DOPA** is broken down outside the CNS, and thus proportionally more is synthesized into dopamine at the sites where it is needed.

(b) Adverse effects of L-DOPA therapy

These include nausea and vomiting, which occur most frequently as the correct dosage of **L-DOPA** is being established. Tolerance to this frequently occurs, but some patients are unable to tolerate this therapy and the drug has to be discontinued. Tolerance also eventually develops to the anti-Parkinsonian effects of **L-DOPA** (presumably as a result of further degeneration of the nigro-neostriatal dopamine neurones). Thus, the dosage of **L-DOPA** may have to be adjusted. Depression, confusion and occasionally psychotic episodes occur in some patients. People who have suffered from psychoses are particularly liable to have a psychotic response following **L-DOPA**; thus they should only be given the drug under closely controlled conditions. It is probable that all these side effects of **L-DOPA** are due to increased dopaminergic function at other sites in the CNS.

Hypotension (orthostatic) can occasionally occur during **L-DOPA** therapy. The reasons for this are not well established, but it has been suggested that **L-DOPA** may increase noradrenaline synthesis at some sites in the CNS and thus cause a decrease in peripheral nervous system sympathetic tone by mechanisms similar to those described for α-methyl-**DOPA** (section 3.6.1). Patients taking **L-DOPA** must not take any preparations such as proprietary tonics which contain pyridoxal phosphate (vitamin B_6). Vitamin B_6 is a cofactor for the enzyme DOPA decarboxylase, and high levels of this vitamin speed the decarboxylation of **L-DOPA** (especially in the liver), with the result that less is available for decarboxylation in the CNS. The beneficial effects of **L-DOPA** are thus reduced.

High doses of **L-DOPA** can cause abnormal involuntary movements (dyskinesias), especially of the tongue and face and dystonic movements of the limbs. The appearance of these movement disorders is dose-dependent and can sometimes be prevented by reducing the dose of **L-**

DOPA. For this reason, it is believed that this type of dyskinesia is caused by excessive stimulation of central dopamine receptors. Other types of dyskinesia are discussed in relation to neuroleptic drugs (section 4.6.4). Occasionally during **L-DOPA** therapy, there are periods during the day when the drug is not effective, followed by periods when control is re-established. These swings in performance are often referred to as 'freezing', or the 'on-off' effect, and may be related to fluctuations in plasma levels of **L-DOPA**, although in many patients the two phenomena are difficult to correlate.

Occasionally the nausea induced by **L-DOPA** therapy can be controlled by concurrent administration of **domperidone**, a selective D2 receptor blocker with anti-emetic properties, which does not cross the blood–brain barrier (section 4.6.4). Likewise some **L-DOPA** induced dyskinetic movements or psychotic episodes can be alleviated by concurrent administration of *low* doses of **pimozide**, a dopamine receptor blocker with some selectivity for D2 sites. In both instances there is a risk of blocking the anti-Parkinsonian actions of **L-DOPA**, or indeed exacerbating the symptoms of Parkinson's disease.

(c) Neurotransmitters in control of motor activity

It has been proposed that the functions of the striatum are to select and maintain motor and other activities initiated in the cortex, which has

---

Figure 4.4 Simplified wiring diagram of the striatum showing pathways which are damaged in Parkinson's disease and Huntington's disease. (a) The circuit consists of the cortical glutamate inputs to the intrinsic ACh and striato-nigral GABA neurones. The activity of both is modulated by the dopamine input from the substantia nigra, pars compacta. The striato-nigral GABA pathway inhibits the GABA pathways to substantia nigra pars compacta and the ventrolateral thalamic nucleus, thus the dopamine pathway from pars compacta and the excitatory pathways from thalamus to cortex are disinhibited. In the undamaged system, this allows expression of striatal function, namely regulation and maintenance of motor activity. (b) In Parkinson's disease, the intrinsic ACh pathway and the glutamate pathway work unopposed, as the dopamine pathway has degenerated. This leads to an inability to initiate movement, or modify movement once it has started, because the cortico/striatal/substantia nigra/thalamic/cortical circuit is no longer modulated by the dopamine loop. (c) In Huntington's disease, degeneration of the striato-nigral GABA pathway means that the GABA pathways from pars reticulata to pars compacta and thalamus are unregulated, so they inhibit the dopamine nigro-striatal pathway and the thalamo-cortical pathway in an unpredictable way. This leads to bursts of involuntary but essentially normally sequenced motor activity. ○—<+, represents an excitatory, ●—<− represents an inhibitory input. Open neurones represent excitatory pathways; closed neurones, inhibitory pathways.

Nigro striatal dopamine pathway

Thalamo cortical
pathway

Cortico striatal pathway
(glutamate)

Intrinsic ACh
neurones

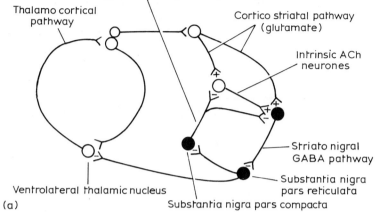

Striato nigral
GABA pathway

Substantia nigra
pars reticulata

Ventrolateral thalamic nucleus

Substantia nigra pars compacta

(a)

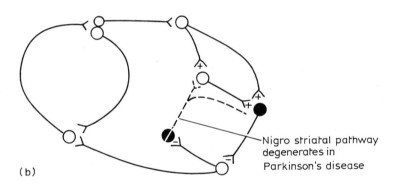

Nigro striatal pathway
degenerates in
Parkinson's disease

(b)

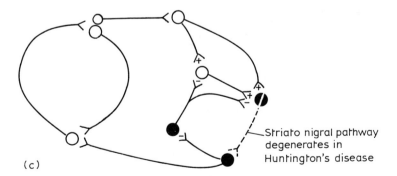

Striato nigral pathway
degenerates in
Huntington's disease

(c)

parallel outputs directly to motor pathways in the spinal cord, and to the striatum.

The roles of dopamine and acetylcholine in regulation of motor activity in the striatum appear to be antagonistic; dopamine is generally inhibitory; ACh is generally excitatory on output neurones. In addition the roles of glutamate (excitatory) and GABA (inhibitory) must be considered.

The striatum has a glutamate input from the cortex which terminates on intrinsic ACh neurones and on GABA neurones which project to the pars reticulata of the substantia nigra (PRSN). There are GABA projections from the pars reticulata to the dopamine cells in the pars compacta of the substantia nigra (PCSN). The dopamine cells project to the striatum ending on the intrinsic ACh neurones. This is illustrated in Fig. 4.4.

In Parkinson's disease, the inhibitory dopamine neurones from PCSN degenerate, thus the excitatory action of ACh and glutamate are unopposed. The GABA pathway to PRSN is continually active and this inhibits the GABA pathways from PRSN to the thalamus and PCSN. The thalamic outputs cease to be inhibited, thus the thalamo-cortico-thalamic loop is unregulated. As the dopamine neurones of the PCSN have degenerated, the GABA input from PCSN is ineffective as it cannot regulate release of a transmitter which is not there. The effectiveness of **L-DOPA** and dopamine agonists in Parkinson's disease appears to be due to inhibitory effects of dopamine on the ACh and GABA neurones in the striatum.

The effectiveness of muscarinic receptor blockers in Parkinson's disease may be ascribed to inhibition of the excess activity of disinhibited intrinsic ACh neurones at the level of the GABA nerve-cell body in the striatum. In each case the drugs serve to normalize the GABA output from the striatum and so allow an inhibitory input from the PRSN to the thalamus.

In Huntington's disease, there is degeneration of the GABA pathway from the striatum to PRCN. Loss of the inhibitory GABA output results in uncontrolled bursts of chorea (which have been likened to inappropriate bursts of normal movement), probably due to intermittent and ineffective inhibition of the thalamo-cortical circuit.

(d) Other drugs used in the treatment of Parkinson's disease

The earliest known effective treatment for Parkinson's disease was the use of muscarinic receptor blockers, which enter the CNS (for example, **atropine** and **hyoscine**). Numerous such compounds are still in use (often in conjunction with **L-DOPA** therapy); they are of most value in the control of tremor and rigidity, but are of little use in alleviating the akinesia. **Benztropine, benzhexol, orphenadrine** and **methixene** are among the more useful anticholinergic drugs for therapy of Parkinson's disease, possibly because, in addition to their anti-muscarinic actions, they also show some blocking effects on the re-uptake of dopamine into nerve terminals, thus making dopamine available for a longer time in the synaptic cleft. Selective blocking of muscarinic receptors in the CNS is not possible, and thus side effects of treatment with such drugs include dry mouth, constipation, blurred vision, glaucoma, urinary retention and other peripheral consequences of muscarinic-receptor blockade (sections 2.4.1 and 2.4.7).

L-DOPA may be considered to stimulate dopamine receptors indirectly. Indirectly-acting dopamine agonists such as the **amphetamines** have only moderate beneficial effects, since their therapeutic actions depend upon the release of endogenously available dopamine. They require intact and functional dopamine neurones and are of limited value in this degenerative disorder. The use of directly-acting dopamine agonists (for example, **apomorphine, bromocriptine** and **piribedil**) in Parkinson's disease is considered in section 4.6.4.

Amantadine, originally introduced as an antiviral agent for influenza, has been found to be effective in a proportion of Parkinsonian patients in controlling rigidity, tremor and akinesia. The mechanisms by which amantadine acts have not been fully established, although evidence for the enhancement of dopamine synthesis and release and the blocking of dopamine re-uptake suggest that facilitation of dopaminergic mechanisms may be involved. It can be given with **L-DOPA**. Side effects include dizziness, insomnia and nausea.

(e) Dopamine foetal cell grafting

Recent studies in man show that foetal dopamine secreting neurones can be successfully grafted into human putamen and result in a marked improvement of motor function in advanced Parkinson's disease.

4.6.2 Disruption of storage. Reserpine and tetrabenazine

Reserpine and **tetrabenazine** disrupt stores of dopamine, and prevent newly synthesized dopamine from being stored within the neurone. **Reserpine** is occasionally used in the treatment of hypertension where central and peripheral noradrenaline mechanisms are believed to be involved (section 3.6.2). A Parkinsonian syndrome can occur, because of decrease in the dopamine available for release. Depression is also a feature of treatment with **reserpine**, but at present there is no general agreement as to which (if any) of the brain amine transmitters – dopamine, noradrenaline or 5-hydroxytryptamine – whose storage is disrupted by

reserpine are normally essentially active to prevent depression being manifested.

Following **reserpine**, hyperprolactinaemia occasionally occurs, and this may be associated with decreased dopamine release in the tubero-infundibular system. Possibly related to this hyperprolactinaemia is the increase in breast cancer in women following long-term therapy with **reserpine**. Animal studies have shown that high circulating prolactin levels are associated with breast cancer.

Tetrabenazine and **reserpine** may be used in the treatment of Huntingdon's chorea, a condition in which there occur abnormal choreiform movements of the limbs, and which, by loss of intrinsic GABA neurones, has been associated with excessive activity of the neostriatal dopamine neurones. The decrease in dopamine in the neostriatum following **tetrabenazine** administration is believed to bring about the decreased incidence of choreiform movements. **Pimozide** and other neuroleptics can be helpful in controlling dyskinesias of Huntington's disease.

4.6.3 Releasers of dopamine. Amphetamines

(+)-**Amphetamine** is a potent dopamine-releasing substance. Many studies have shown that at doses lower than those needed to activate other neurochemical systems, **amphetamine** releases dopamine, at the same time blocking its re-uptake. The ability of **amphetamine** to cause dopamine release is in agreement with present theories on the mechanisms of some forms of schizophrenia (especially paranoid psychosis – section 4.6.4), which are believed to be caused by overactivity of the mesolimbic dopaminergic system. In man, high doses of (+)-**amphetamine** are associated with the appearance of a psychotic condition which is not distinguishable from paranoid psychosis with auditory hallucinations and persecution-directed ideas. **Amphetamine**-induced psychosis has been used in volunteers as a model for schizophrenia. **Amphetamine** intoxication and psychosis can be effectively treated with neuroleptics.

Lower doses of **amphetamine** produce both mental and physical arousal and this is the basis of its stimulant actions. It is characterized by an increased capacity for both physical and mental activity, a delay in the onset of fatigue and decreased need for sleep. This is followed by a period of 'hangover' during which depression can be a prominent feature. This after-effect makes **amphetamines** unsuitable for the treatment of endogenous or exogenous depression, and **amphetamines** are therefore not classified as antidepressants. Increasing doses can lead to a state of euphoria, and higher doses still can induce the psychotic syndrome.

Amphetamine-like compounds have been used as anoretics; drugs used to decrease appetite for food. They are now rarely used as their adverse effects, especially dependence, and liability for abuse for recreational purposes, creates problems more difficult to resolve than the obesity which they are designed to treat.

Amphetamines are sometimes used in the treatment of narcolepsy, which is a condition in which people fall into a light sleep from which they can readily be aroused. This occurs during waking hours, and increased stimulation by **amphetamine** can enable such people to remain awake. (+)-**Amphetamine** and **methylphenidate** have been used in the treatment of the 'hyperkinetic syndrome' in children. It is not clear how any benefit is derived by the children.

4.6.4 Dopamine-receptor agonists and blockers

(a) Dopamine and dopamine-receptor agonists

Dopamine itself is not given orally as it is destroyed in the intestines and liver by MAO and COMT. **Dopamine** is occasionally given by intravenous infusion in the treatment of shock. The effectiveness of **dopamine** in shock probably depends on the increased cardiac output (dopamine has weak β-adrenoceptor agonist activity), and the alteration in the balance of vasoconstriction and vasodilation towards increased peripheral vasoconstriction (because **dopamine** has some agonist actions at α-adrenoceptors mediating vasoconstriction). Importantly however **dopamine** causes dilatation of the renal arteries with subsequent increased blood flow to the kidneys, and thereby reduces the risk of kidney failure due to inadequate blood supply in cardiogenic, septicaemic, and other forms of shock.

Apomorphine has been used to cause vomiting in cases of poisoning. The chemoreceptor trigger zone in the area postrema of the medulla oblongata is involved in the control of nausea and vomiting. Many chemical stimulants if applied to this site (or *in vivo* if carried to this site in the bloodstream), will induce vomiting. This site is also sensitive to the actions of dopaminergic and cholinergic agonists which induce vomiting, and also to the action of dopaminergic and cholinergic receptor blockers, which can act as anti-emetics. **Apomorphine** is a direct dopamine-receptor agonist. As **L-DOPA, apomorphine** and **bromocriptine** cause nausea and vomiting, there is reason to believe that dopaminergic mechanisms are involved in the control of these functions.

Apomorphine is currently being reviewed as a therapy in Parkinson's disease. As the drug is not absorbed in reasonable quantity from the gastrointestinal tract it must be administered systemically. Current treatments give this drug subcutaneously either as a bolus injection or by continuous infusion from a pump strapped to the body. To avoid its potent emetic action, **domperidone** must also invariably be given.

Bromocriptine is a direct dopamine receptor agonist which has been found to be of value in the treatment of neuro-endocrine disorders, namely hyperprolactinaemia (which commonly results in galactorrhoea, excessive milk secretion), amenorrhoea (absence of menstruation) and infertility and acromegaly. **Bromocriptine** is believed to act on the dopamine receptors in the anterior pituitary which control prolactin release. Dopamine is believed to be prolactin inhibitory hormone (PIH), which normally is released from nerve terminals in the hypothalamic median eminence, and is carried to the anterior pituitary gland by the portal system, where it inhibits prolactin release. In certain pathological states, it would seem that there is insufficient dopaminergic nerve activity in the median eminence, which results in insufficient amounts of dopamine (PIH) being released into the portal blood vessels. With insufficient PIH, excess amounts of prolactin are released from the anterior pituitary. Prolactin acts on the mammary glands, causing galactorrhoea; acting in the hypothalamus, it inhibits release of GnHRH (gonadotropic-hormone releasing hormone). In the absence of FSH (follicle-stimulating hormone) and LH (luteinizing hormone), ovulation in the female and spermatogenesis in the male cease, thus resulting in

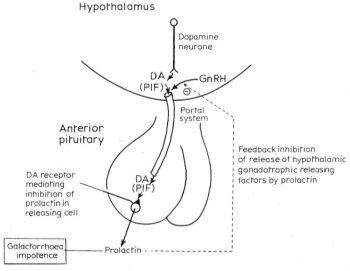

Figure 4.5 Control of prolactin release by dopamine (prolactin inhibitory hormone).

infertility. By acting directly on the dopamine receptors in the anterior pituitary, **bromocriptine** mimics the actions of dopamine (PIH); thus, prolactin release is decreased to normal levels, galactorrhoea no longer occurs and regular menstruation and fertility are restored (Fig. 4.5). **Bromocriptine** is also used to inhibit puerperal lactation.

Whereas it is not clearly established that **dopamine** acts in a similar manner in the control of growth hormone (somatotrophic-hormone) release, the efficacy of **bromocriptine** in the treatment of acromegaly, a pathological condition in which there is excess secretion of growth hormone, suggests that a similar mechanism might be involved. **Bromocriptine** reduces circulating levels of growth hormone and this leads to a reversal of the symptoms of acromegaly.

Bromocriptine, lysuride and **pergolide** are used in the treatment of Parkinson's disease, and these drugs can be effective in patients not adequately responding to, or not tolerating, **L-DOPA** therapy. The main side effects of these dopamine receptor agonists are nausea, abdominal upsets, postural hypotension and drug induced dyskinesias.

(b) Dopamine-receptor blockers. Drugs and treatment of psychoses

The terms neuroleptic, antipsychotic or major tranquilliser are applied to drugs used therapeutically in the treatment of psychoses. These include the dopamine-receptor blockers. Once the terms included the Rauwolfia alkaloids (e.g. **reserpine**), which block dopaminergic neurotransmission by disruption of storage, but these are no longer used as antipsychotic drugs.

Schizophrenia is a collective term used to describe several clinical conditions which are characterized by symptoms including: (1) thought disorder (in which people think and talk in an incomprehensible way); (2) decreased and/or inappropriate emotional responses; (3) auditory hallucinations (in which people hear voices giving them orders and saying uncomplimentary things about them); (4) delusions of being persecuted (paranoia); and (5) withdrawal from social contact. Other symptoms (such as depression, mania and anxiety) may occur concurrently with those which are characteristic of schizophrenic psychosis, and may require treatment.

Chlorpromazine was the first effective antipsychotic and it is still extensively used. **Fluphenazine** is a phenothiazine but, in additon to the orally administered form, **fluphenazine decanoate** and **fluphenazine enanthate** can be given by deep intramuscular (depot) injections, from which the active compound is slowly released over a period of 2–4 weeks, thus allowing the drug to be administered periodically on an out-patient basis. As lack of patient compliance is liable to be a major consideration in schizophrenia when treatment outside hospital is being contemplated, depot injections of neuroleptics are of great value in long-term management. Some of the numerous neuroleptic drugs used in the treatment of schizophrenia include the butyrophenones such as **haloperidol** and **trifluperidol**; thioxanthines such as **thiothixene** and **flupenthixol**; and **pimozide, thioridazine** and **sulpiride**.

It is argued that those dopamine receptor blockers which show greater affinity to the D2 receptor are better at controlling the symptoms of schizophrenia as it is suggested that they exhibit less extrapyramidal side effects. However in clinical practice neuroleptics which block both D1 and D2 receptors are equally effective as antipsychotic drugs. Many of these drugs are also sedative in their action which is beneficial in managing the agitated patient or those with behavioural disturbances are characterized with manic features.

The initial association of neuroleptic activity and dopamine receptor blockade was made from the observation that patients treated with neuroleptics developed the symptoms of Parkinson's disease (section 4.6.1). Subsequent biochemical and pharmacological studies have given rise to the dopamine theory of schizophrenia, in which it is suggested that overactivity of some parts of the dopaminergic system in the CNS (particularly the mesolimbic dopamine system) results in schizophrenic symptoms. Thus, drugs which block dopamine receptors relieve schizophrenia, but at the same time their action in the nigro-striatal dopamine system gives rise to the symptoms of Parkinson's disease and other extrapyramidal side effects. The dopamine theory in schizophrenia is compatible with the observations of psychotic episodes occurring following the administration of drugs such as (+)-**amphetamine**, which cause dopamine release (section 4.6.3).

Direct biochemical evidence for abnormal dopaminergic function in schizophrenia has come from measurements made in post mortem brain in schizophrenic patients. When comparison was made with aged matched controls, high concentrations of dopamine were found in the left amygdala of brains of schizophrenic patients. This contrasted with symmetrical distributions of dopamine concentrations in the left and right amygdala of brains from non-schizophrenic people. Ligand binding studies showed that there is an increase in the number of D2 binding sites in the left amygdala from brains of schizophrenic patients. The nature of the dopamine abnormalities in the brains of schizophrenic patients must be considered alongside the neurological observations that damage to the left temporal lobe (in which the amygdala is found) has a high association with psychotic behaviour.

Drugs which block dopamine receptors can be of value as anti-emetics. Thus, phenothiazines and butyrophenones can usefully control drug-induced vomiting. They cannot be used to treat nausea and vomiting following **L-DOPA**, since the beneficial anti-Parkinsonian effect of **L-DOPA** is also abolished. **Metoclopramide** blocks some actions of dopamine, especially nausea and vomiting, but it has no antipsychotic activity. It is used only as an anti-emetic, although extrapyramidal symptoms can develop. **Sulpiride** is also of value as an anti-emetic, and the drug is claimed to possess antipsychotic properties. **Domperidone** is an anti-emetic agent which principally acts by blocking D2 receptors in the chemoreceptor trigger zone (area postrema) of the medulla. It relieves nausea and vomiting associated with cytotoxic chemotherapy. It does not appreciably cross the blood–brain barrier and therefore is claimed not to be sedative or induce extrapyramidal reactions. This property is useful in controlling the nausea associated with **L-DOPA** therapy and of dopamine agonists (**bromocriptine, lysuride, apomorphine**) used in the treatment of Parkinson's disease.

Neuroleptics also have some minor tranquillizing (anxiolytic) actions, and they are sometimes used (at low doses) just for their sedative actions; **chlorpromazine** and **thioridazine** are examples. This sedation by neuroleptics is sometimes made use of in the management of terminal pain, as the decrease in awareness of pain brought about by opiate analgesics is potentiated. The mechanism of this potentiation is not understood. Neuroleptanalgesia is a form of analgesia sometimes used in surgery when the patient's co-operation is required, and it is achieved by combining opiate analgesics with neuroleptics.

The ability of neuroleptics to cause dopamine release (by an action at presynaptic receptors) might also account for the reported effectiveness of these compounds in the treatment of depression. The role of individual monoamine neurotransmitters in endogenous depression is not established, and findings that low doses of the phenothiazines have antidepressant action suggest that dopaminergic mechanisms might be involved (section 4.6.5). In addition, the phenothiazines (e.g. **chlorpromazine**) have found therapeutic applications in a number of pathological states, including Huntington's chorea (section 7.1.6), anorexia nervosa (section 4.6.3), vertigo, mania and alcoholism.

(c) Side effects of dopamine receptor blockers

Neuroleptics can induce a Parkinsonian syndrome by blocking dopamine receptors of the nigro-neostriatal pathway. It is suggested that those neuroleptics with marked muscarinic-receptor blocking action (e.g. **clozapine**), do not cause Parkinsonian symptoms as often as those neuroleptics which are relatively free of anticholinergic action (e.g. **haloperidol** and **trifluperidol**). Antimuscarinic drugs are used in the treatment of Parkinson's disease (section 4.6.1), and can reduce tremor and rigidity of the Parkinsonian syndrome induced during treatment with neuroleptics.

In addition to extrapyramidal symptoms, neuroleptic therapy can lead to disturbances in neuro-endocrine regulation. The galactorrhoea, infertility and amenorrhoea which may occur during treatment with neuroleptic drugs are probably related to blockade of dopamine receptors in the anterior pituitary (see the remarks on bromocriptine in section 4.6.4). The excess prolactin released by such actions is believed to be associated with the increased incidence of breast cancer which occurred in women taking neuroleptics and **reserpine** (once used as a major tranquillizer). Neuroleptic treatment often leads to increased feeding and obesity (see the remarks on **amphetamine**-induced anorexia in section 4.6.3).

Pigmentation of the skin due to deposits of melanin sometimes occurs following long-term administration of phenothiazines; if this occurs on the cornea it can impair sight. Agranulocytosis is sometimes reported following neuroleptic therapy.

The antimuscarinic action of neuroleptics (e.g. phenothiazines and thioxanthenes) can cause blurred vision, dry mouth and constipation, as well as other unwanted effects associated with this action (section 2.4.7). Phenothiazines in general, and **chlorpromazine** in particular, also have some α-adrenoceptor blocking actions, which might account for postural hypotension. The phenothiazines were originally discovered during a search for better antihistaminic drugs. It is thus not surprising that many of them possess histamine receptor blocking properties, which may, in part, account for the drowsiness and sedation they cause (section 6.6.4). **Chlorpromazine** can cause epileptiform seizures in susceptible individuals.

(d) Neuroleptic-induced dyskinesias

In addition to the Parkinsonian syndrome which may occur following administration of neuroleptics, other extrapyramidal symptoms may develop acutely, including motor restlessness (akathesia) and acute dyskinesia (usually characterized by dystonic postures of the neck, trunk and limbs). These acute dyskinetic reactions can usually be controlled by drugs with anticholinergic properties (e.g. **benztropine** and **promethazine**) or by benzodiazepines (e.g. **diazepam**). It is believed that they are caused by an imbalance of cholinergic and dopaminergic activity in extrapyramidal brain regions following the administration of the neuroleptic drug. In addition to their dopamine receptor blocking actions, the neuroleptics are believed to stimulate dopamine release by acting at pre-synaptic dopamine receptors (section 4.4.3); such actions may result in the postulated cholinergic:dopaminergic imbalance.

A further form of extrapyramidal movement disorder can occur following years of neuroleptic use, namely the tardive dyskinesias. These involuntary movements usually take the form of oro-facial movements and/or dystonic posturing of the limbs and trunk. The tardive dyskinesias do not respond to anticholinergic drugs, which in fact exacerbate the symptoms. Paradoxically, the movements can be effectively controlled by increasing the dose of neuroleptic drug. However the treatment is not resolute as the movements are likely to re-occur at the higher dose level. These observations suggest that the tardive dyskinesias occur when dopamine receptors become supersensitive to dopamine following chronic blockade. Chronic neuroleptic treatment will enhance pre-synaptic dopaminergic mechanisms (through presynaptic receptors, section 4.4.3), which may eventually overcome postsynaptic receptor blockade. As in denervation supersensitivity (section 1.9.4), the response to receptor stimulation of the drug-induced supersensitive postsynaptic receptor may be greater than the normal maximum, causing involuntary movements.

Indeed experiments have convincingly shown that chronic treatment of animals with dopamine receptor blocking drugs for 1–6 months will both markedly increase receptor binding affinity and also the number of receptor sites. This drug-induced supersensitivity is believed to be the mechanism of the chronic tardive dyskinetic reaction seen in man.

(e) Neuroleptic malignant syndrome

Very rarely the acute administration of an antipsychotic drug to a naive patient can precipitate an array of side effects including catatonia, hyperpyrexia and changing level of consciousness as well as autonomic disturbances such as sweating and cardiovascular collapse characterized by tachycardia and hypotensive and hypertensive episodes. This rare complication of drug treatment is called the neuroleptic malignant syndrome and in a fulminant state is life-threatening requiring cardiovascular, respiratory and nutritional support until the effect of the drug has worn off, a state which may last several days, especially if it has been precipitated by a depot injection of neuroleptic.

The actions of dopamine receptor agonists and receptor blockers used therapeutically are summarized in Table 4.2.

Table 4.2 Summary of drugs which modify dopaminergic transmission

Mechanism	Drug	Effect	Uses
Synthesis	**L-DOPA**	Increased synthesis	Parkinson's disease
	α-Methyl-_p_-tyrosine	Inhibits tyrosine hydroxylase	Experimental
	Carbidopa **Benserazide**	Inhibits DOPA decarboxylase outside CNS	Parkinson's disease (with **L-DOPA**)
Storage	**Reserpine** **Tetrabenazine**	Disrupt dopamine storage	Major tranquillizers (obsolete)
	MAO inhibitors	Enhanced dopamine storage	
Release	**(+)-Amphetamine** **Mazindol** **Tyramine**	Release dopamine onto receptors	Anorectics, CNS stimulants
Receptors	**Dopamine** **Apomorphine** **Bromocriptine** **ADTN** **Pirbedil**	Non-selective dopamine receptor agonists	Emetic Parkinson's disease Prolactin-induced subfertility Acromegaly
	Fenoldepam SKF38393	Selective D1 agonists	
	SCH23390	Selective D2 blocker	Experimental

	Drug	Action	Use
	Quinpirole	Selective D2 agonist	
	Chlorpromazine **Fluphenazine** **Haloperidol** **Pimozide**	Non-selective D1 and D2 Dopamine receptor blockers	Major tranquillizer, antipsychotic schizophrenia, mania, anti-emetic
	Metoclopramide **Sulpiride**	Selective D2 blockers	
Inactivation of uptake	**Domperidone** **Amphetamines** **Cocaine** **Nomifensine**	Inhibitors of dopamine uptake	Anorectics, CNS stimulants
	Benztropine **Benzhexol** **Mazindol**		Parkinson's disease
of metabolism	**Iproniazid** **Tranylcypromine** **Phenelzine**	MAO inhibitors non-selective	Antidepressants
	Selegiline	Selective for MAO B	Parkinson's disease

4.6.5 Inactivation of released dopamine

(a) Inhibitors of dopamine re-uptake

Substances which block the re-uptake of dopamine into presynaptic nerve terminals will prolong the effect of released dopamine in the synaptic cleft. Drugs which have such an action include **benztropine, orphenadrine** and **benzhexol**, which also block muscarinic receptors. These compounds are used in the treatment of Parkinson's disease, as they facilitate the effects of **L-DOPA**, and their anticholinergic actions are of value in the control of tremor and rigidity (section 4.6.1).

Tricyclic antidepressants (sections 3.5.1 and 3.6.5) have only relatively weak dopamine uptake inhibitory actions; this cannot be discounted in their possible mechanism(s) of action as antidepressants. It is possible that lack of dopaminergic activity occurs in endogenous depression, in which lack of initiation of motor activities is a major feature. Certain MAO inhibitors, for example, **tranylcypromine**, also inhibit dopamine re-uptake. This might contribute to the antidepressant actions of this compound (section 3.6.5).

(b) MAO inhibitors

These also arise in discussing noradrenaline (section 3.6.5) and 5-hydroxytryptamine (section 5.6.5). Dopamine is destroyed intracellularly by MAO and concentrations of dopamine increase following inhibition of this enzyme by drugs such as **iproniazid, nialamide, isocarboxazid** and **tranylcypromine**. Whether increased concentrations of dopamine contribute towards the antidepressant action of MAO inhibitors is not established. **Selegiline** has recently been tested as a selective inhibitor of MAO B, the iso-enzyme of MAO which has a particuarly high substrate affinity for dopamine. **Selegiline** is claimed to be a useful adjunct to **L-DOPA** therapy in some patients with Parkinson's disease, as it enhances and prolongs this drug's action. It is also suggested that **selegiline** may have a protective action on nigro-striatal neurones, slowing the rate of degeneration in this disorder, and therefore delaying the rate of progression of the disease.

If treatment with **selegiline** is started at diagnosis, it is claimed that a longer time elapses before the need to begin treatment with **L-DOPA** when compared with appropriate controls who have not been given **selegiline**. This observation has been taken to indicate that inhibition of MAO B may prevent the production of a toxic metabolite which is responsible for the degenerative process. The identity of such a neurotoxic compound is as yet unknown, but the ability of MAO B inhibitors to prevent the neurotoxicity of MPTP is used as evidence to support such a hypothesis.

FURTHER READING

Kebabian, J.W. and Calne, D.B. (1979) Multiple dopamine receptors. *Nature*, 277, 93–6.

Kopin, I.J. and Markey, S.P. (1988) MPTP toxicity: Implication for research in Parkinson's disease. *Ann. Res. Neurosci.*, 11, 81–96.

Langston, J.W. and Irwin, I. (1986) MPTP current concepts and controversies. *Clin. Neuropharmacol.*, 9, 485–507.

Lees, A.J. (1986) L-DOPA treatment and Parkinson's disease. *Q. J. Med.*, 59, 535–47.

McGeer, E.G., Staines, W.A. and McGeer, P.L. (1984) Neurotransmitters in the basal ganglia. *Can. J. Neurolog. Sci.*, 11, suppl., 89–99.

Marsden, C.D. (1990) Parkinson's disease. *Lancet*, 948–52.

Penney, J.B. and Young, A.B. (1983) Speculations on the functional anatomy of basal ganglia disorders. *Ann. Rev. Neurosci.*, 6, 73–94.

Sokoloff, P., Giros, B., Martes, M-P., Bouthenet, M-L. and Schwartz, J-C. (1990) Molecular cloning and characterization of a novel dopamine receptor (D3) as a target for neuroleptics. *Nature*, 347, 146–51.

Tanner, C.M. (1989) The role of environmental toxins in the aetiology of Parkinson's disease. *Trends Neurosci.*, 12, No. 2.

5 5-Hydroxytryptamine

5-Hydroxytryptamine (5-HT, serotonin) is a neurotransmitter in the central nervous system (CNS) and in the myenteric plexus of the gut. High concentrations are found in the enterochromafin cells in the lining of the alimentary canal and in the blood platelets. 5-HT is released from platelets at sites of tissue injury, and 5-HT in the systemic circulation (plasma) has a wide range of effects on nerve terminals and blood vessels which do not have a 5-HT innervation. The physiology and pharmacology of 5-HT is complex, for in addition to being a neuro-transmitter it has local hormone or paracrine functions.

5-HT is a monoamine, and many features of its synthesis, storage, release and inactivation are similar to the processes occurring in tissues which synthesize the other monoamines, namely noradrenaline, dopamine and adrenaline.

5.1 SYNTHESIS

5-HT is made from the aromatic amino acid L-tryptophan, which is hydroxylated to 5-hydroxytryptophan (5-HTP) by the enzyme tryptophan hydroxylase. Tryptophan hydroxylase is found only in the cytoplasm of 5-HT neurones and is the rate-limiting step enzyme in the synthesis of 5-HT (Fig. 5.1). This enzyme requires molecular oxygen and reduced pteridine cofactor for activity. L-Tryptophan is actively taken up into 5-HT neurones by means of a neutral amino acid carrier mechanism, which itself may be a limiting step in the synthesis of 5-HT (section 5.1.1). 5-HTP is decarboxylated in the cytoplasm to 5-HT by the non-specific enzyme L-aromatic amino acid decarboxylase (5-HTP decarboxylase).

5.1.1 Control of 5-HT synthesis

Under normal physiological conditions tryptophan hydroxylase is not fully saturated, i.e. the enzyme is not working to full capacity. Therefore,

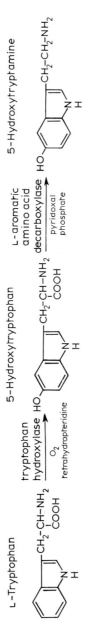

Figure 5.1 Synthesis of 5-HT.

unlike its counterpart in catecholamine neurones (tyrosine hydroxylase, which utilizes tyrosine), tryptophan hydroxylase activity can be influenced by the amount of free L-tryptophan available in the presynaptic nerve terminal. The amount of available L-tryptophan is dependent on a number of factors including the concentration of circulating L-tryptophan in the plasma and the rate of its uptake into the brain and presynaptic nerve terminals. Administration of **L-tryptophan** in depression and schizophrenia increases plasma tryptophan concentrations and results in increased synthesis of 5-HT in the CNS.

The relationship between the rate of 5-HT synthesis in the CNS and the plasma levels of L-tryptophan is not simple. L-Tryptophan is transported in plasma either in solution or bound to plasma proteins. The rate of entry of L-tryptophan into the brain depends upon the ratio of free : bound tryptophan in the plasma, and this ratio is influenced by the concentration in the blood of neutral amino acids and drugs, which compete for the plasma protein binding sites, as well as for the tryptophan-uptake sites. 5-HTP is taken up by neurones other than just 5-HT neurones; therefore the increases in 5-HT synthesis are not selectively confined to 5-HT neurones.

5.1.2 Inhibitors of enzymes of 5-HT synthesis

These are only of experimental value. Irreversible tryptophan hydroxylase inhibitors include **DL-parachlorophenylalanine, 6-fluoro-tryptophan** and α-propyldopacetamide. Inhibitors of 5-HTP decarboxylase include hydrazine derivatives (e.g. **carbidopa**) and α-methyl-5-HTP.

5.2 STORAGE

The 5-HT storage mechanisms have many features in common with catecholamine storage processes (section 3.2). 5-HT is believed to be bound in a complex with proteins, divalent ions and adenosine triphosphate. 5-HT is actively taken up from the cell cytoplasm into these granules.

5.2.1 Disruption of 5-HT storage

Reserpine and **tetrabenazine** disrupt the granular storage of 5-HT, causing it to leak into the cytoplasm, where it is metabolized intraneuronally by monoamine oxidase. Some 5-HT might also leak extraneuronally. The amounts of stored 5-HT are markedly reduced following use of **reserpine** or **tetrabenazine**.

5.3 RELEASE

Experimental evidence suggests that neuronal 5-HT can be released into the synaptic cleft by the process of exocytosis in response to action potentials and to drugs. This release process is dependent upon an influx of calcium into the neurone.

5.3.1 Facilitation of 5-HT release

Drugs which release 5-HT from tryptaminergic neurones include (+)-**amphetamine**, and **fenfluramine**). The halogenated amphetamines (e.g. **fenfluramine** and **parachloramphetamine**) are more potent than the non-halogenated amphetamines at releasing 5-HT. Some tricyclic antidepressants, e.g. **chlorimipramine** (**clomipramine**) and **amitriptyline**, can release 5-HT, as well as block its neuronal re-uptake (the latter action is also exerted by **fenfluramine** and **parachloramphetamine** (section 5.5.1). The 5-HT-releasing and re-uptake blocking actions of these drugs may contribute to their use as antidepressants (**chlorimipramine** and **amitriptyline**)) and anorectics (**fenfluramine**).

5.3.2 Regulation of 5-HT release

Release of 5-HT from neurones is regulated by axon traffic, availability of precursor **L-tryptophan**, and by auto- and heteroreceptors. Inhibitory 5-HT autoreceptors are found on somatodendritic and axon terminal sites. Release of 5-HT from chromafin granules is sensitive to vagal stimulation (possibly by a cholinergic heteroreceptor) and mechanical stimuli. Release from platelets occurs in response to tissue injury, both mechanical and chemical.

5.4 5-HT RECEPTORS

The first classification of 5-HT receptors (which is now largely of historic interest) was proposed by Gaddum and Picarelli. They proposed the existence of two types of 5-HT receptor which they designated D and M. The D receptor was so called because the contraction of intestinal smooth muscle by 5-HT could be competitively blocked by dibenyline; (dibenyline is a trade name for **phenoxybenzamine**, which is also a non-selective α-adrenoceptor blocker). The M receptor was so called because **morphine** could block the actions of 5-HT at parasympathetic nerve endings where it caused release of acetylcholine.

The present day classification of 5-HT receptors recognizes three receptor sub-types. It is based on functional studies involving selective agonists and selective and specific antagonists, radio-ligand binding studies, identification of receptor genes and studies of transducer and effector mechanisms. The three main groups have been designated 5-HT1, 5-HT2 and 5-HT3; the 5-HT2 receptors correspond to the D receptor of the original classification and the 5-HT3 corresponds to the M receptor.

5.4.1 5-HT1 receptors

Four sub-types of 5-HT1 receptors (designated 1a, 1b, 1c, 1d) are proposed although it is likely that alterations to this scheme will occur for reasons outline below. The 5-HT 1a receptor is best characterized; the gene codes for a 421 amino acid receptor protein linked by G protein(s) to two effector systems, namely inhibition of adenylyl cyclase and opening of a G protein sensitive K^+ channel. 5-HT 1a receptors are found at somatodendritic and axon terminal sites where they act as 5-HT autoreceptors. 5-HT 1a receptors are also found postsynaptically in the CNS. 8-OH-DPAT (**8-hydroxydipropyl amino tetraline**) is a selective agonist at 5-HT 1a receptors; **ipsapirone** and **buspirone** are selective agonists or partial agonists at 5-HT 1a receptor sites. In common with other 5-HT1 receptors there are no selective 5-HT 1a antagonists.

The 5-HT 1b and the 5-HT 1d receptors can be considered together. The gene for these receptors has not been isolated but they appear to be G protein regulated receptors as their activation leads to inhibition of adenylyl cyclase. No selective agonists or antagonists are available for 5-HT 1b or 5-HT 1d receptors. The distinction between 5-HT 1b and 5-HT 1d receptors is based on studies of non-selective 5-HT agonists and blockers; they are both autoreceptors at axon terminals. The 1b receptor is unique to the rat (in which species most research work has been done), the 1d receptor is found in other mammalian species including man.

The inclusion of the 5-HT 1c receptor among the 5-HT1 receptors is controversial as G protein transduction mechanisms and agonist and antagonist studies suggest that it has more features in common with 5-HT2 receptor sites than with true 5-HT1 sites; the suggestion has been made that there may be sub-types of the 5-HT2 receptor to which the 5-HT 1c receptor should be added.

A 5-HT 1c receptor gene has been isolated and it codes for a 460 amino acid G protein regulated structure which activates phospholipase C leading to formation of the secondary messengers inositol trisphosphate and diacyl glycerol. α-**Methyl 5-HT** is a selective agonist at 5-HT 1c receptors; **retanserine** and **pizotifen** are selective antagonists for this receptor. The similarities between this 5-HT 1c receptor and the 5-HT2 receptor considered below are remarkable.

The sub-division of 5-HT1 receptors described above does not accommodate a group of receptors which have been called 5-HT1-like receptors, which are found postsynaptically in some intestinal and vascular smooth muscle. With the exception of 5-HT c sites all 5-HT1 and all 5-HT1-like receptors are selectively stimulated by **5-carboxamidotryptamine**. **Methiothepin** is a potent antagonist at these receptors but it is not selective in that it also blocks 5-HT2 receptors. **Sumitriptan** (GR43176) is a selective 5-HT1-like receptor agonist which does not readily fall into the 5-HT1 sub-type classification. From the above description it may be appreciated that there is still much to discover about the physiology and pharmacology of 5-HT systems.

5.4.2 5-HT2 receptors

A gene coding for the 5-HT2 receptors has been identified; it codes for a 471 amino acid receptor protein linked by a G protein to phospholipase C, activation of which leads to synthesis of secondary messengers inositol trisphosphate and diacyl glycerol. α-Methyl 5-HT is a selective agonist at 5-HT2 receptors, **retanserine, ketanserine** and **pizotifen** are selective blockers.

5.4.3 5-HT3 receptors

The gene for the 5-HT3 receptor has not been identified. Preliminary evidence suggests that 5-HT3 receptors form part of an ion channel for sodium, potassium and calcium ions. 5-HT3 receptors have been identified in the autonomic nervous system and in the CNS. 5-HT3 receptors are found on many nerve terminals where they regulate the release of neurotransmitter; they appear to constitute therefore an important group of heteroreceptors. In the autonomic nervous system they release noradrenaline from sympathetic nerves and acetylcholine from parasympathetic nerves. In the CNS they may facilitate dopamine release at mesolimbic axon terminals.

Information about the existence and functional roles of 5-HT3 receptors has been derived largely from studies of potent and selective antagonists. **Metoclopramide** and **cocaine** are weak 5-HT3 receptor blockers; **odansetron** (GR38032F), ICS205930 and MDL72222 are potent 5-HT3 receptor blockers. **2-methyl-5-HT** and **phenylbiguanid** are selective 5-HT3 receptor agonists.

5.4.4 5-HT neurones

Recent findings show that two anatomically distinct types of 5-HT axons are found in the CNS. Axons arising from the dorsal raphe nucleus are very fine with small varicosities which innervate limbic structures and the striatum; axons from the median raphe nucleus have a larger diameter and contain large varicosities, and innervate the hippocampus and mammillary body. The significance of the two classes of axon terminals is not clear but they show differential sensitivity to the actions of neurotoxins and they may prove to be the targets of selectively acting drugs as synthesis, storage, release and re-uptake mechanisms in the two types of axon terminal appear to be different; whether the distribution of pre- and postsynaptic receptors and regulation of release is different in these axons are questions which await answers.

5.4.5 5-HT in the CNS

The distribution of 5-HT in the CNS forms a diffuse network, and exact functional roles are not firmly established. The nerve-cell bodies of the major 5-HT neurones are in the midline raphe nuclei of the rostral pons (Fig. 5.2). Ascending fibres innervate the basal ganglia, hypothalamus, thalamus, hippocampus, limbic forebrain and areas of the cerebral cortex. 5-HT-containing cell bodies in the brain stem give rise to descending axons which terminate in the medulla and the spinal cord. Indirect evidence from experiments and observations in animals and man suggest that central 5-HT mechanisms are important in the control of mood and behaviour, motor activity and its control, feeding and control

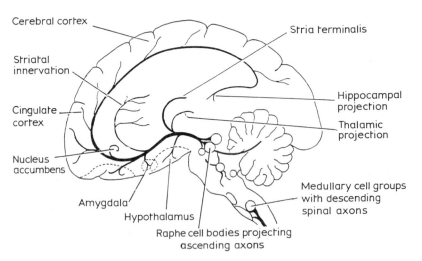

Figure 5.2 5-HT pathways in the CNS. The nerve-cell bodies are found in the raphe nuclei and in the medulla; the distribution of 5-HT axons is illustrated.

of hunger, thermoregulation, sleep, certain hallucinatory states, and possibly some neuro-endocrine control mechanisms in the hypothalamus.

5.5 INACTIVATION OF 5-HT

5.5.1 Neuronal uptake of 5-HT

A high-affinity, energy-dependent, active-transport mechanism exists to remove 5-HT from the synaptic cleft back into the presynaptic neurone. Once inside the presynaptic nerve terminal, 5-HT will be further taken back up into the storage vesicles by another high-affinity, energy-dependent transport system.

(a) Inhibitors of neuronal uptake of 5-HT

These include the tricyclic antidepressants (**imipramine, desimipramine, amitriptyline** and **chlorimipramine**) and the anorectic agent **fenfluramine**. The effect of these drugs is to make more neurotransmitter available in the synaptic cleft and thus effectively prolong the actions of postsynaptic receptor stimulation. The tricyclic antidepressants also block the neuronal uptake system for noradrenaline (section 3.6.5, where their mode of action in the alleviation of the symptoms of depression has been considered in more detail). The halogenated derivative of **imipramine** – **chlorimipramine** (**clomipramine**) – is more potent at blocking 5-HT uptake than noradrenaline uptake. **Fluoxetine** and **citalopram** are highly selective inhibitors of 5-HT uptake into neurones.

5,6-Dihydroxytryptamine and 5,7-dihydroxytryptamine are neurotoxins which are selectively taken up by 5-HT neurones by the active-transport process. They cause degeneration of 5-HT neurones and are used experimentally for tracing 5-HT pathways and investigating 5-HT functions.

5.5.2 Enzymatic destruction of 5-HT. MAO

Any 5-HT not bound into storage vesicles will be converted into inactive metabolites by the enzyme monoamine oxidase (MAO). MAO is a mitochondrially located enzyme which oxidatively deaminates 5-HT into 5-hydroxyindole acetaldehyde; this is then converted by the actions of an aldehyde dehydrogenase into 5-hydroxyindoleacetic acid (5-HIAA), the major inactive metabolite of 5-HT (Fig. 5.3). 5-HT is a substrate for MAO type A.

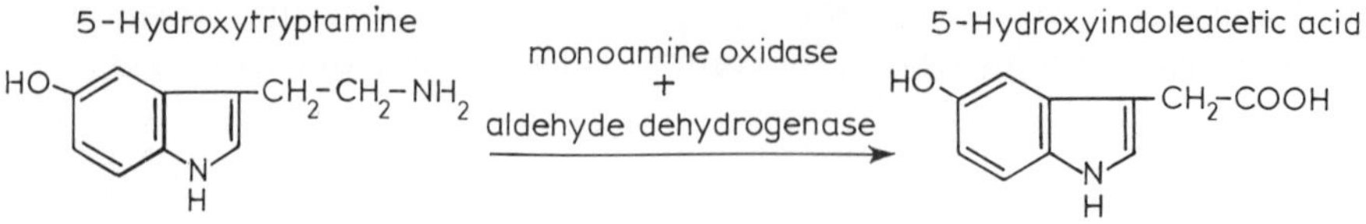

5-Hydroxytryptamine

$$HO-\text{[indole ring]}-CH_2-CH_2-NH_2$$

monoamine oxidase
+
aldehyde dehydrogenase
⟶

5-Hydroxyindoleacetic acid

$$HO-\text{[indole ring]}-CH_2-COOH$$

Figure 5.3 Metabolism of 5-HT.

Inhibitors of MAO

In addition to their capacity to prevent the enzymatic breakdown of the catecholamines (sections 3.5.2 and 4.5.2), MAO inhibitors also block the oxidative deamination of 5-HT. Treatment with such compounds leads to increased levels of 5-HT in enterochromaffin granules and within nerve terminals in the CNS. Presumably more 5-HT is available for release. MAO inhibitors used clinically as antidepressants include **phenelzine, tranylcypromine, nialamide, iproniazid** and **isocarboxazid**.

If MAO is inhibited, 5-HT is metabolized to N-methyl, or N,N-dimethyl or O-methyl tryptamine by hydroxy-N-methyltransferase or hydroxyindole O-methyltransferase. High concentrations of these enzymes are found in the pineal gland, where they are important in the synthesis of melatonin (N-acetyl-5-methoxytryptamine).

5.6 THERAPEUTIC APPLICATIONS AND CONSEQUENCES OF DRUGS ACTING AT 5-HT SYNAPSES

5.6.1 Drugs which affect 5-HT synthesis. L-Tryptophan

L-Tryptophan acts as a precursor of 5-HT synthesis, and administration of **L-tryptophan** to man has been found beneficial in some depressive states. It has been suggested that the plasma concentration of free tryptophan is decreased in depressive illness; although total tryptophan levels are not lower, it is suggested that a larger percentage is bound to plasma proteins (section 5.1.1). Administration of **L-tryptophan** increases free plasma tryptophan concentrations, brain and neuronal uptake of tryptophan, and also the neuronal and extraneuronal (peripheral) synthesis of 5-HT. In states of depression, **L-tryptophan** can be given either alone or in conjuction with a MAO inhibitor, the effect of the latter being to prolong the life of the newly synthesized 5-HT by blocking its metabolism. The actual increase in 5-HT synthesis after **L-tryptophan** will be determined by a number of factors including diet and the presence of other drugs which alter the ratio of free : bound tryptophan in the plasma. In addition, uptake of **L-tryptophan** into the brain and into 5-HT neurones can be influenced by the concentrations of other aromatic amino acids, such as phenylalanine and tyrosine, which compete for the same carrier site.

The adverse side effects of **L-tryptophan** administration are usually associated with elevated 5-HT function at both central and peripheral sites. Symptoms include sedation, prolongation of sleep, nausea, blurred vision and diarrhoea; following high doses of **L-tryptophan**, carcinoma of the bladder has been reported. When given in conjunction with a MAO inhibitor, adverse side effects of tremor, hyperreflexia, flushing, orthostatic hypotension and nystagmus have been reported. Some of the unwanted peripheral effects of **L-tryptophan** can be controlled with the 5-HT receptor blockers **cyproheptadine** and **methysergide**.

5.6.2 Drugs which affect 5-HT storage. Reserpine

In addition to disrupting the storage of catecholamines, **reserpine** and **tetrabenazine** cause a marked reduction in the stores of 5-HT both in the CNS and peripherally. In the CNS it is believed that most of the 5-HT whose storage is disrupted by **reserpine** is inactivated by MAO, and that this might be associated with the sedation and depression which occur following the administration of **reserpine**. This depression can occasionally lead to suicide. In the gastro-intestinal tract, it appears that, when **reserpine** disrupts 5-HT storage, considerable quantities of 5-HT are released to act on 5-HT receptors; this gives rise to side effects such as increased gut motility, abdominal cramps and diarrhoea.

5.6.3 Drugs which affect 5-HT release. Fenfluramine

Fenfluramine is used clinically as an anorectic, and it causes release of 5-HT. The anorectic action is believed to be related to the stimulation of 5-HT receptors in the CNS. That 5-HT mechanisms might be involved in the control of feeding is further indicated by the appetite-stimulant action of a 5-HT-receptor blocker – **cyproheptadine**. **Fenfluramine** can cause mood disturbances, especially depression, when the drug is discontinued; thus, it should not be given to persons with a history of depressive illness. Peripheral side effects of **fenfluramine** are relatively few, but the diarrhoea which occurs is probably related to 5-HT release in the gastro-intestinal tract.

5.6.4 Functional roles of 5-HT1 and 5-HT1-like receptors

5-HT1 and 5-HT1-like receptors are found both pre- and postsynaptically in the peripheral and in the CNS. Presynaptic 5-HT1a receptors function as somatodendritic inhibitory autoreceptors in the raphe nuclei and in other brain stem 5-HT nuclei. At axon terminals 5-HT 1a receptors act as inhibitory autoreceptors, regulating release of 5-HT. 5-HT1-like receptors are also found on sympathetic axon terminals where they act to inhibit release of noradrenaline. The precise physiological role of 5-HT1 and 5-HT1-like receptors will not be adequately defined until selective antagonists of these receptors become available.

(a) 5-HT and migraine

Release of 5-HT (either from nerves or from platelets) occurs during migraine attacks. 5-HT causes constriction of all large blood vessels and following the constrictor phase (prodrome or warning phase of migraine) during which there are visual disturbances owing to constriction of retinal blood vessels, there follows a prolonged phase of vasodilation during which characteristic throbbing headache and nausea are prominent features. It has been suggested that the dilator phase is caused by insufficient release of 5-HT (from nerves or platelets) and dilation of blood vessels occurs as a result of release of prostaglandins and other vasodilators. The dilator phase has been likened to a sterile inflammation.

The above description of the complex role of 5-HT and other mediators of migraine allows a tentative explanation of the apparently contradictory efficacy of 5-HT agonists and blockers in the treatment of migraine. Drugs with weak and non-selective 5-HT1 and 5-HT1-like receptor blocking properties and partial agonists at 5-HT1 and 5-HT1-like receptors, for example **methysergide**, might be effective by preventing the vasoconstrictor phase and drugs with partial agonist activity might be effective by their ability to constrict blood vessels at a time when there is insufficient endogenous 5-HT available. The 5-HT1-like agonist **sumitriptan** may be effective by virtue of its ability to prevent the vasodilator phase. The usefulness of anti-inflammatory analgesics such as aspirin and paracetamol in migraine may be due to their prevention of the synthesis of prostaglandins; they are inflammatory mediators which sensitize pain receptors and which cause vasodilation. The role (if any) of 5-HT2 receptors in migraine is not established. **Ergotamine**, a partial agonist at 5-HT2 and α-adrenoceptors is still available for treatment of migraine although unpredictable responsiveness and adverse effects such as gangrene of the fingers due to vasoconstriction of peripheral blood vessels makes it a potentially dangerous drug.

Stimulation of 5-HT2 receptors leads to contraction of all smooth muscle. The source of 5-HT leading to constriction of bronchi, gut and blood vessels in response to 5-HT is not known; neuronal, platelet derived and chromaffin cell derived 5-HT are likely sources. Selective 5-HT2 antagonists **retanserin** and **ketanserin** have no established therapeutic uses although they have been tried in the treatment of migraine and hypertension.

Pizotifen is useful in the prophylactic treatment of migraine, but can cause sedation due to its antihistamine properties. **Methysergide** can also be used to control severe recurrent migraine and cluster headaches, but it may not be used for longer than 6 months in succession due to dangers of development of retroperitoneal fibrosis.

The importance of 5-HT2 receptors in the CNS is evidenced by experimentation in animals and in man. 5-HT2 receptors have a role in sensory processing; **lysergic acid diethylamide** (LSD) is a non-selective antagonist at 5-HT2 receptors and **bufotenin** and **N,N-dimethyl 5-HT** are hallucinogens with partial agonist activity at 5-HT2 receptors.

A disorder of 5-HT mechanisms is strongly indicated in depressive illness. The effectiveness of antidepressives which block re-uptake of 5-HT (**clomipramine, fluoxetine, citalopram**) and the effectiveness of **L-tryptophan** in treatment of depressive illness, suggest a role for 5-HT in this condition. Prolonged treatment with antidepressives and in some studies ECT leads to a decrease in the number (down regulation) of 5-HT2 binding sites in the CNS. The functional significance of this down regulation is not understood as it does not occur with all antidepressives. An involvement of 5-HT2 dependent mechanisms in cognition, short term memory and other higher cerebral functions is indicated by the decrease in the number of 5-HT2 binding sites in dementia patients.

(b) 5-HT3 receptors

5-HT3 receptors are found on sympathetic and parasympathetic nerve endings in the peripheral nervous system. Stimulation leads to release of acetylcholine from parasympathetic nerve endings and noradrenaline from sympathetic nerve endings. The complex cardiovascular effects consequent on intravenous injection of 5-HT are known as the Betzold–Jarish reflex. The selective and competitive 5-HT3 receptor blockers **odansetron** (GR38032F), MDL72222, and ICS205-930 selectively and competitively block the Betzold–Jarish reflex.

In the CNS 5-HT3 antagonist binding sites have been found concentrated in the area postrema of the brainstem, in the cerebral cortex and in the mesolimbic system. 5-HT3 receptor blockers are highly effective anti-emetic agents which are able to control nausea and vomiting caused by nearly all chemotherapeutic agents and radiation; **odansetron** is now used for this purpose. In addition clinical trials are under way to establish whether animal data which suggest that 5-HT3 receptor blockers may have antipsychotic and anxiolytic action, can be confirmed in man. If confirmed, the antipsychotic activity may be linked to the ability of these compounds to inhibit the release of dopamine in the mesolimbic system. Studies in animals suggest that 5-HT acting at 5-HT3 receptors can facilitate dopamine release. Were it to prove possible to regulate enhanced dopamine function (which is believed to exist in schizophrenia and other psychotic states), then use of 5-HT3 receptor blockers may be anticipated to exert an antipsychotic effect, but without extrapyramidal adverse effects.

The results of basic and clinical studies with 5-HT3 receptor blockers and their possible effectiveness in the treatment of anxiety are awaited.

Numerous animal models of anxiety states indicate that enhancement of 5-HT transmission has an anxiogenic action whilst decreased 5-HT function has an anxiolytic effect. 5-HT 1a agonists, for example 8-OH-DPAT, are active in some of these models. Selective stimulation of somatodendritic or axon terminal 5-HT inhibitory autoreceptors would lead to decreased 5-HT transmission and if such a mechanism indeed operates then it would be consistent with the hypotheses that reduced 5-HT function reduces anxiety. **Buspirone** has recently been introduced as an anxiolytic for clinical use; detailed reports of its efficacy in man are awaited.

(c) 5-HT and pain

Part of the descending 5-HT pathways of the raphe nuclei and lower brain stem nuclei terminate in the dorsal horn of the spinal cord. Fibres carrying sensory information from peripheral structures terminate in the dorsal horn of the spinal cord and there is a dense 5-HT innervation at this site. It has been suggested that 5-HT may regulate the transmission of painful stimuli at the level of the dorsal horn and at brain stem sites especially in the periaquaducal grey matter. Drugs which enhance 5-HT transmission, for example blockers of re-uptake (**zimelidine, clomipramine**) may have some analgesic actions in their own right or may enhance the action of opiate analgesics. A definitive statement as to the role of 5HT (if any) in regulation of pain transmission, the mechanism(s), and sites of any such action await clarification (section 9.4.5).

The role of 5-HT in regulation of sleep is controversial; speculation as to the possible role of 5-HT in sleep mechanisms arose from experiments in cats in which destruction of the raphe nuclei caused insomnia. Experiments with existing selective and specific 5-HT agonists and blockers fail to support any simple and general theory linking 5-HT with regulation of sleep.

5-HT1 mechanisms are associated with central regulation of

autonomic function such as blood pressure, thermoregulation and eating and sexual behaviour.

Fenfluramine is occasionally used in the treatment of obesity as it has an anorectic effect, which means it reduces food intake. It has been suggested that **fenfluramine** releases 5-HT which works at a postsynaptic site where it activates mechanisms of satiety possibly in the hypothalamus.

5.6.5 Inactivation of 5-HT

(a) Drugs which inhibit neuronal uptake of 5-HT

The main class of compounds effective at inhibiting the neuronal uptake of 5-HT comprises the tricyclic antidepressants. Some of the indirectly acting 5-HT receptor agonists (e.g. **fenfluramine**) also block the neuronal uptake of 5-HT, and their main pharmacological actions are believed to be related to inhibition of 5-HT uptake and release from presynaptic terminals.

The tricyclic antidepressants, such as **amitriptyline, nortriptyline, imipramine, desimipramine, chlorimipramine (clomipramine)**, are administered for the alleviation of the symptoms of depression (depression is discussed in detail elsewhere; section 3.6.5). Tricyclic antidepressant drugs block the neuronal re-uptake of both noradrenaline and 5-HT, with weaker effects on dopamine uptake. The halogenated derivative **clomipramine** appears to be more selective at blocking 5-HT uptake than **noradrenaline**. The side effects of these drugs are related to their anticholinergic properties – this accounts for such side effects as dry mouth, constipation, blurred vision and urinary retention – and to their blocking actions on α-adrenoceptors and H_1-histamine receptors. The numerous and clinically important interactions of the tricyclic anti-depressants with other drugs are considered in section 3.6.5.

(b) MAO inhibitors

Administration of MAO inhibitors results in elevation of cerebral 5-HT concentrations by blocking the major route of 5-HT metabolism. Such compounds are useful in the treatment of depression. Clinically used MAO inhibitors include **tranylcypromine, phenelzine, iproniazid, nialamide** and **isocarboxazid**. In addition to blocking 5-HT metabolism, MAO inhibitors also inhibit noradrenaline and dopamine metabolism, with subsequent increase in the concentrations of catecholamines. The major side effects of these drugs are related to their actions on the sympathetic nervous system. These, together with the important

Table 5.1 Summary of drugs which modify 5–Hydroxytryptaminergic transmission

Mechanism	Drug	Effect	Uses
Synthesis	**L-Tryptophan** **p-Chlorophenylalanine**	Increases synthesis Inhibits tryptophan hydroxylase	Antidepressant Experimental, carcinoid syndrome
Storage	**Reserpine** **Tetrabenazine** MAO inhibitors	Disrupt 5-HT storage Enhance 5-HT storage	Antidepressants
Release	Fenfluramine Chlorphentermine Amphetamines	Release 5-HT onto receptors	Anorectics
Receptors	**5-HT** **LSD** **DMT** **Quipazine** **8-OH-DPAT** Ipsapirone α-methyl-5-HT α-methyl-5-HT 2-methyl-5-HT	Non-selective 5-HT-receptor agonists Selective 5-HT la agonists Selective 5-HT lc agonist Selective 5-HT2 agonist Selective 5-HT3 agonist	Hallucinogens Anxiolytic

Drug	Mechanism	Use
5-carboxamidotryptamine GR43175-somitriptan **Methiothepin**	Selective 5-HTI-like agonists Non-selective blocker at 5-HTI-like receptors	Antimigraine
Ketanserin **Pizotifen** **Ritanserin**	Selective 5-HT 1c blockers Selective 5-HT2 blockers	
GR38032F (odnasetron) ICS 205930 MDL 72222	Selective 5-HT3 blockers	Anti-emetic
Citalopram **Fluoxetine** Tricyclic antidepressants	Inhibitors of neuronal 5-HT uptake	Antidepressants
Fenfluramine		Anorectic
Phenelzine **Iproniazid** **Tranyecypromine**	MAO inhibitors	Antidepressants

Inactivation of uptake

of metabolism

LSD, lysergic acid diethylamide; DMT, dimethyltryptamine; 8-OH-DPAT, 8-hydroxydiproyl amino tetraline.

considerations of the interaction of MAO inhibitors with other therapeutic agents, are presented in section 3.6.5.

A summary of drugs which modify 5-hydroxytryptaminergic transmission is given in Table 5.1.

FURTHER READING

Andrews, P.L.R., Rapeport, W.G. and Sanger, G.J. (1988) Neuropharmacology of emesis induced by anti-cancer therapy. *TIPS*, **9**, No. 9.

Cooper, S.J. (1989) Drugs interacting with 5-HT systems show promise for treatment of eating disorders. *TIPS*, **10**, No. 2.

Fozard, J.R. and Gray, J.A. (1986) 5-HT_{1c} receptor activation: a key step in the initiation of migraine ? *TIPS*, **10**, No. 8.

Hartig, P.R. (1986) Molecular biology of 5-HT receptors. *TIPS*, **10**, No. 2.

Hoyer, D. and Middlemiss, D.N. (1989) Species difference in the pharmacology of terminal 5-HT autoreceptors in mammalian brain. *TIPS*, **10**, No. 4.

Peroutka, S.J. (1988) 5-Hydroxytryptamine receptor subtypes: molecular, biochemical and physiological characterization. *TINS*, **11**, No. 11.

Saxena, P.R. and Ferrari, M.D. (1989) 5-HT_1-like receptor agonists and the pathophysiology of migraine. *TIPS*, **10**, No. 8.

Tricklebank, M.D. (1989) Interactions between dopamine and 5-HT_3 receptors suggest new treatment for psychosis and drug addiction. *TIPS*, **10**, No. 4.

6 Histamine

Histamine is a neurotransmitter in the central nervous system (CNS). Outside the CNS, histamine is found in mast cells, basophils and histaminocytes in the stomach. Its functional roles in the CNS are not well established; in the stomach it is an important stimulus for the secretion of hydrochloric acid; its release from mast cells and basophils occurs following allergic reactions and injury. Histamine is an important regulator of repair processes in damaged tissues.

6.1 SYNTHESIS

Histamine is formed by the decarboxylation of L-histidine to L-histamine by the enzyme histidine decarboxylase, which uses pyridoxal phosphate (vitamin B_6) as cofactor (Fig. 6.1). The non-specific enzyme L-aromatic amino acid decarboxylase can also catalyse this reaction. L-Histidine is a semi-essential amino acid, and at present there are no therapeutic uses made of drugs which inhibit histidine decarboxylase. The availability of L-histidine to histidine decarboxylase is believed to be the rate-limiting step in histamine synthesis. No indications exist for increasing histamine synthesis by administering L-histidine. It is not clearly established whether any histamine made by bacteria in the gut is ever taken up and used; the ability of the liver to inactivate circulating histamine is such that any contribution from this source is a minor one.

High concentrations of histidine decarboxylase are found in nerves and in cells which synthesize and release large quantities of histamine.

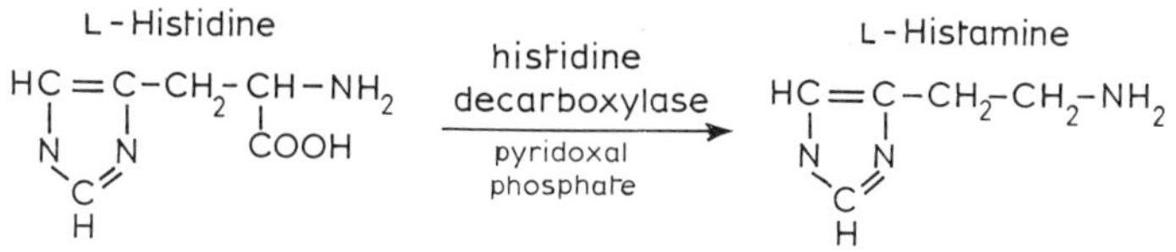

Figure 6.1 Synthesis of histamine.

Antibodies to histidine decarboxylase have been used for immuno-cytochemical mapping of histamine pathways in the CNS.

6.2 STORAGE

Outside the CNS, histamine is stored in mast cells and platelets, and in these cells it is complexed with heparin and ATP. No histamine is found in nerves outside the CNS. It is released from mast cells and platelets in response to stimuli such as tissue injury and antigen–antibody reactions. In the CNS, in gastric mucosa, lungs and skin, histamine is stored in a different complex to that in mast cells, namely in a complex which has some similarities to the monoamine storage complex found in noradrenergic, dopaminergic and 5-hydroxytryptamine-containing neurones (section 3.2).

6.3 RELEASE

6.3.1 Histamine release from mast cells and basophils

Numerous stimuli are known to cause calcium-dependent histamine release. Tissue injury such as trauma, heat and chemicals (whether drugs in therapy or encountered in everyday life) can all cause histamine release: release occurring through cell disruption or through a secretory process. During tissue injury, released bradykinin and substance P can themselves release histamine. Plant poisons (from nettles, for example) and animal venoms (from bee or wasp stings) release histamine as well as other autacoids. Antigen–antibody reactions, which are seen as allergic and anaphylactic responses, are in part mediated by histamine release. The various stimuli of histamine release and their therapeutic considerations are discussed in section 6.6.3.

Histamine release is a calcium-dependent process; the processes are complex and differ between cell type; entry of extracellular calcium and liberation of calcium from intracellular stores are variously important.

6.3.2 Release of histamine from other sites

Compound 48/80, a polymeric macromolecular substance, is used experimentally to release histamine from mast cells but, when its effects are examined on sympathetic nerves, only a small proportion of the histamine associated with these nerves is released. It is at present not established whether histamine can be released from peripheral nerves.

Vagal stimulation and acetylcholine stimulate gastrin release, which by independent actions activates histidine decarboxylase, releases histamine and directly stimulates gastric acid secretion. The mechanisms by which acetylcholine and gastrin (and also synthetic pentagastrin) induce histamine release are at present not established.

6.3.3 Inhibition of release

The ability of compounds to prevent the release of histamine (as well as other mediators of allergic and anaphylactic reactions) has led to their prophylactic use in allergic disorders such as asthma. **Disodium cromoglycate (DSCG)** and **nedocromil** prevent the release of biological mediators of bronchoconstriction (including histamine), but which site in the sequence from antigen–antibody reaction to release of mediator is affected is not clearly defined.

6.4 HISTAMINE RECEPTORS

On the basis of studies with histamine receptor agonists and blocking compounds, three histamine receptors have been identified. The three types, which differ in agonist and antagonist sensitivities, and mediate different physiological responses, have been called histamine H_1, H_2 and H_3 receptors.

6.4.1 Histamine receptors outside the CNS

Dependent on the tissue, histamine has varying effects on smooth muscle. Bronchiolar smooth muscle has H_1 receptors which contract in the presence of histamine to produce bronchoconstriction. This may occur during allergic asthma and in some anaphylactic responses. H_1-receptor blockers are only weakly active in preventing bronchoconstriction, probably because mediators other than histamine are also released. Acting on H_1 receptors, histamine causes contraction of intestinal smooth muscle.

The actions of histamine on vascular smooth muscle are complex, species-dependent and mediated by both H_1 and H_2 receptors. The smooth muscle of arterioles, small veins and the intervening capillaries are relaxed by histamine, while some large arteries and veins are constricted. There is overall relaxation of cerebral blood vessels, and this can cause vascular headache. In man, the overall effect of histamine released into the bloodstream is a fall in blood pressure, which can be serious if baroreceptor reflexes are diminished, as in general anaesthesia.

A further effect of histamine acting at H_1 receptors is to increase the permeability of capillaries, which results in leakage of plasma fluids into the extracellular space and manifests as oedema, rhinitis and conjunctivitis.

(a) The Lewis triple response

This was originally described for the events observed following mild injury caused by pulling a blunt instrument firmly across the skin; it is more reliably seen following intradermal injection of histamine. At the site of injection there appears a red spot (flush), as a result of the dilation of blood capillaries. This lasts for about 1 min, after which it fades and is replaced by a white raised wheal resulting from increased capillary permeability and the consequent oedema. Around the red spot (and later the wheal), there is an area of red flare, which frequently does not appear until the wheal has replaced the red spot. The flare is believed to be due to local dilation of blood vessels under the control of a poorly defined axon reflex (but see section 9.5 substance P). It is not related to the spread of injected histamine. If applied on a dermal blister, histamine can lead to itching and pain; endogenously released histamine can presumably do the same.

(b) Gastric acid secretion

Gastric acid secretion from the parietal cells is stimulated by injected and mucosally released histamine acting at H_2 receptors.

6.4.2 Histamine receptors in the CNS

Histamine has an uneven distribution in the CNS and, whereas terminals containing histamine have been tentatively identified, the nerve-cell bodies have not been detected. No functional roles for histamine in the CNS are firmly established but, on the basis of experience with drugs which block histamine receptors, it has been suggested that histamine might have a role in arousal, in mechanisms related to nausea and vomiting, and in the control of blood pressure and water metabolism. Ligand binding studies have shown H_1, H_2 and H_3 receptors in the CNS.

6.4.3 Histamine receptor sub-types

H_1 receptors were the first sub-type of histamine receptor to be characterized. No selective agonists are known at the H_1receptor, but **mepyramine, diphenhydramine, promethazine** and **chlorpheniramine** are selective competitive blockers of H_1 receptors. H_1 receptors are linked to phospholipase C by a Gs regulatory protein; diacyl glycerol and inositol triphosphate are the secondary messengers generated.

H_2 receptors can be selectively activated by **dimparit**. **Cimetidine** and **ranitidine** are selective competitive blockers of H_2 receptors which are linked to adenylyl cyclase by a Gs regulatory protein.

H_3 receptors have been identified as inhibitory histamine autoreceptors on all cells which release histamine. **R-α-methylhistamine** is a selective agonist, and **thioperamide** is a selective competitive antagonist at H_3 receptors. Neither the gene nor the transducer/effector system for the H_3 receptor have been identified.

6.5 INACTIVATION OF HISTAMINE

Following release, histamine diffuses back into cells, but there does not appear to be a high-affinity uptake system for its removal. Blood-borne histamine is largely metabolized during passage through the liver. There are three main enzymes involved in the degradative metabolism of histamine to pharmacologically inactive products. Histamine-N-methyl transferase adds a methyl group (from S-adenosyl methionine as donor) to yield N-methyl histamine, and this is further degraded by monoamine oxidase (MAO). Alternatively, histamine can be oxidatively deaminated by diamine oxidase (DAO), to give imidazole acetic acid. The precise metabolic pathway varies in different species (Fig. 6.2).

6.6 THERAPEUTIC APPLICATIONS AND CONSEQUENCES OF DRUG ACTION ON HISTAMINE SYSTEMS

6.6.1 Synthesis of histamine

It is not possible to modify histamine synthesis by increasing substrate availability or inhibiting synthesis in a therapeutically useful way.

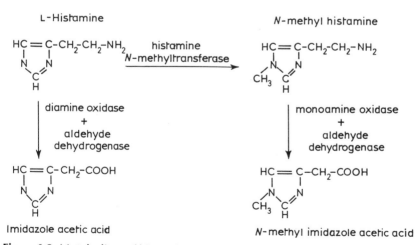

Figure 6.2 Metabolism of histamine.

6.6.2 Storage of histamine

Compounds disrupting histamine storage have not found clinical applications: any possible disruption of histamine storage by **reserpine** is not considered therapeutically important (but see section 6.6.3).

6.6.3 Release. Basic drugs. DSCG

Many forms of tissue injury can cause histamine release. Chemicals and antigens may cause histamine release when they are ingested or inhaled or are in contact with the skin. Such reactions can occur in sensitive individuals to a very large range of substances, such as foreign proteins found in dust, pollen or food, or drugs which form haptens. Following direct injury by trauma, histamine release may be direct, or it may be indirect, through the release of mediators such as bradykinin, substance P, 5-hydroxytryptamine, prostaglandins or slow-reacting substance of anaphylaxis (SRS-A). Nettle stings and also wasp and other insect stings can cause histamine release. Many drugs (especially basic ones) – for example, (+)-**tubocurarine, morphine, dextran, reserpine, codeine** and **pethidine** – and various antibiotics can all evoke histamine release when they are rapidly injected intravenously.

The effects of histamine release may be distributed through many systems in the body, and it is not always possible to predict the main results. However, those systems in direct contact with the causative agent are most likely to be affected. In the respiratory system, broncho-constriction is the most marked response, and the ability to exhale is most severely impaired. In the gastro-intestinal tract, histamine release can lead to excess acid secretion and contribute to gastric ulceration or its maintenance. Spasm of smooth muscle can lead to pain. Histamine released near the surface of the skin can cause urticaria, itching and pain. In man, vasodilation leading to lowering of blood pressure is the most prominent feature of histamine release within the cardiovascular system. This can be precipitated by intravenously injected drugs and ingested poisons. Hay fever and rhinitis commonly occur following the inhalation of antigens which release histamine through immunological mechanisms. In addition to the H_1-receptor antagonists, physiological antagonists of histamines such as the sympathomimetic amines (section 3.4.4) are most effective in the treatment of bronchoconstriction following histamine release.

Inhibition of histamine release

Drugs which prevent the release of histamine (and other mediators), such as **disodium cromoglycate (DSCG)** and **nedocromil** must be given prophylactically, since they are not effective in preventing histamine release once the process of mediator release has started, as in an asthmatic attack. For example, DSCG is delivered to the lungs in the form of a fine powder suspension from a 'spinhaler'. DSCG can be used in hay fever; the powder is sniffed into the nose.

6.6.4 Agonists and blockers at H_1 and H_2 histamine receptors

(a) Histamine-receptor agonists

Histamine and **betazole** have now been superseded by **pentagastrin** in tests made to assess the capacity for gastric acid secretion. When used, **histamine** and **betazole** are injected subcutaneously, and an H_1-receptor blocker is given previously to prevent actions other than those on the gastric mucosa.

(b) H_1-receptor blockers

There are many H_1-receptor blockers used therapeutically, including **promethazine, mepyramine, chlorpheniramine** and **diphenhydramine**. They can be of value in the treatment of hay fever and similar conditions affecting the eyes and nose. Urticaria and itching following stings can be successfully treated with H_1-receptor blockers. H_1 blockers can be of value in the therapy of migraine, but they are of limited use in the treatment of broncho-constriction. Sedation is a common side effect.

    Terfenidine and **astemizole** are H_1 blockers which do not readily cross the blood–brain barrier, and so only rarely cause drowsiness and sedation. They are used in the symptomatic treatment of allergic rhinitis, hay fever and urticaria.

    Cinnarizine is a H_1 blocker which causes little sedation, but which appears to enter the CNS as it is effective in the treatment of motion sickness, Menières disease and vertigo. **Promethazine** and **mepyramine** are also used for these indications, but sedation usually occurs.

    H_1-receptor blockers are used in anaesthetic premedication to cause sedation and also for their anticholinergic actions. Individuals vary markedly in their sensitivity to the sedative effects of H_1 blockers and, in some persons (especially children), a stimulant response may occur. Given in the evening, H_1 blockers can act as hypnotics. Overdose can lead to coma and death. In addition to blocking H_1 receptors, these compounds can to varying degrees also block cholinergic (muscarinic) receptors and thus exhibit many anticholinergic side effects. **Promethazine** has marked anticholinergic side effects (section 2.4.6), and its use as an anti-emetic in motion sickness and in Parkinson's disease to control tremor and rigidity is related to its anticholinergic actions rather than to any actions it might have on cerebral H_1 receptors. Phenothiazines (e.g. **chlorpromazine**) and some anticholinergic compounds (e.g. **atropine**) also have marked H_1-receptor blocking properties. H_1-receptor blockers are usually given orally, but for rapid onset of action they can be given intramuscularly or intravenously.

(c) H_2 histamine receptor blockers

Cimetidine was the first clinically used H_2 histamine receptor blocker, and it was found to be effective in treatment both of gastric and duodenal ulcers, for it reduced the acidity of the stomach (and duodenal) contents and therefore facilitated ulcer healing. Following some years of use, some side effects have been observed with **cimetidine**, especially in the elderly (in whom dosage should be altered), and this has been ascribed to deteriorating liver and kidney function. Confusion has been one of the

Table 6.1 Summary of drugs which modify histaminergic mechanisms

Mechanism	Drug	Effect	Uses
Synthesis	—	—	
Storage	—	—	
Release	Injury		
	(+)-**Tubocurarine**		
	Morphine	Histamine release	
	Dextran	onto receptors	
	Disodium cromoglycate	Inhibits histamine	Asthma
	Nedocromil	release	
Receptors	**Histamine**		
	2-Methyl histamine	H_1 activation	
	Mepyramine		
	Promethazine	H_1-histamine-	Allergy, travel
	Chlorpheniramine	receptor blockers	sickness
	Terfenidine		
	Betazole	H_2 activation	Gastric acid
	Dimparit		secretion
	Cimetidine		
	Burimamide	H_2 blockers	Gastric and
	Metiamide		duodenal
	Rinatidine		ulcers
	R-α-methylhistamine	Selective H_3 agonist	
	Thioperamide	H_3 blockers	

adverse effects reported, and subfertility due to excess secretion of prolactin (section 4.6.4). These two observations suggest that **cimetidine** may enter the CNS and that it can directly or indirectly influence anterior pituitary function. More recently **ranitidine** has been introduced as an alternative H_2 histamine receptor blocker.

6.6.5 Inactivation of released histamine

The relation of the probable increase in tissue histamine concentrations following treatment with MAO inhibitors to the therapeutic effects of these compounds is not known.

A summary of drugs which modify histaminergic mechanisms is given in Table 6.1.

FURTHER READING

Black, J.W., Duncan, W.A.M., Durant, C.J., Ganellin, C.R. and Parsons, E.M. (1972) Definition and antagonism of histamine H_2 receptors. *Nature*, **236**, 385–90.

Hirschowitz, B.I. (1979) H_2 histamine receptors. *Ann. Rev. Pharmacol.*, **19**, 203–44.

Prell, G.D. and Green, J.P. (1986) Histamine as a neuroregulator. *Ann. Rev. Neurosci.*, **9**, 209–54.

Van der Werf, J.R. and Timmerman, H. (1989) The histamine H_3 receptor: a general presynaptic histaminergic regulatory system? *TIPS*, **10**, No. 4.

7 Inhibitory amino acids: GABA, glycine and taurine

7.1 AMINO ACIDS AS NEUROTRANSMITTERS

Certain amino acids found within the mammalian central nervous system (CNS) appear to fulfil many of the criteria used to define a neurotransmitter (Chapter 1). Amino acids found in the peripheral nervous system do not conform to all the criteria, although it is possible that some serve a neurotransmitter role at autonomic ganglia. Amino acids within the brain and spinal cord have a major metabolic in addition to a neurotransmitter role. As well as being incorporated into proteins (with the exception of GABA), the amino acid neurotransmitters are closely linked to the metabolism of glucose, one of the major substrates for energy metabolism in the nervous system. Separation of metabolic and neurotransmitter function is often very difficult; indeed direct measurement of amino acid concentrations in brain regions can be a poor indicator of neurotransmitter function.

Although the distribution of neurotransmitter amino acids is often ubiquitous within the CNS, the main sites of pharmacological action are small interneurones, although long projecting pathways have been described (e.g. strio-nigral GABA neurones). Immunohistochemical and receptor binding techniques have allowed detailed mapping of proposed amino acid neuronal pathways, with identification of transmitter synthesizing enzymes and presynaptic storage vesicles, and location of amino acid receptor binding sites.

From electrophysiological studies, the proposed amino acid neurotransmitters may be considered in two main functional classes: the excitatory acidic amino acids, containing two carboxylic acid groups (L-glutamic acid, L-aspartic acid), and the monocarboxylic or inhibitory neutral amino acids (γ-aminobutryic acid, glycine, taurine, proline, β-alanine).

Excitatory amino acid neurotransmitters open Na^+ channels and so increase Na^+ conductance; this leads to depolarization and increased neuronal firing rate. Inhibitory (or depressant) amino acid neurotransmitters open Cl^- channels, cause increased chloride conductance of postsynaptic membranes, and cause hyperpolarization and decreased neuronal firing rate.

Other amino acids which have been considered as possible neurotransmitters include L-cysteic acid, L-cysteine sulphonic acid, L-homocysteic acid and L-homocysteine sulphonic acid (as excitatory amino acids) and L-α-alanine, hypotaurine, L-serine and L-cystathionine (as inhibitory amino acids). Evidence for these compounds having a true neurotransmitter role is often equivocal. A point of difficulty with certain amino acids is that they have potent actions at many neurones within the CNS which are not adjacent to presynaptic nerve terminals releasing that particular amino acid. Thus amino acids cannot be considered to have a neurotransmitter role at all sites at which they show biological activity.

At present, few clinical states are firmly associated with malfunction of amino acid neurotransmission, although in the last few years a number of new exciting developments have occurred which may soon change this situation.

7.2 γ-AMINOBUTYRIC ACID

γ-Amino-*n*-butyric acid (GABA) is an inhibitory neurotransmitter in the brain, spinal cord and retina. It is not present in peripheral nerves, although trace amounts of GABA have been identified in some autonomic ganglia, where its functional role, if any, remains to be established.

7.2.1 Synthesis

GABA is formed by the decarboxylation of L-glutamic acid by the enzyme glutamic acid decarboxylase (GAD). GAD is found exclusively in the cytoplasm of presynaptic GABA nerve terminals and, like other amino acid decarboxylases, GAD requires pyridoxal phosphate (vitamin B_6) as cofactor (Fig. 7.1.).

(a) Control of GABA synthesis

GAD is probably the rate-limiting step in GABA synthesis. However it is not totally certain if GABA can influence GAD activity by the process of end-product inhibition. Saturating concentrations of L-glutamic acid are

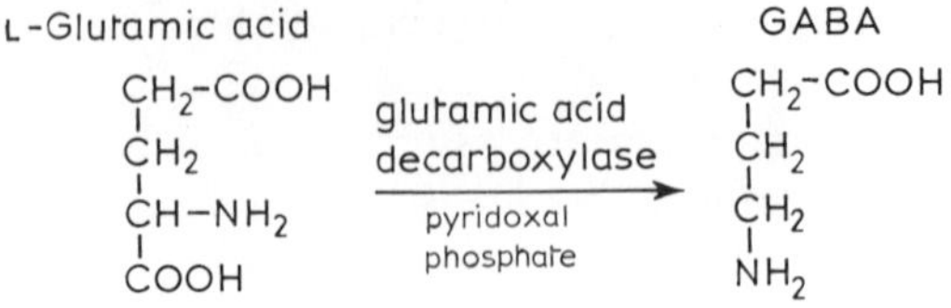

L-Glutamic acid   GABA

CH$_2$-COOH   glutamic acid   CH$_2$-COOH
CH$_2$ decarboxylase   CH$_2$
CH-NH$_2$ ⟶   CH$_2$
COOH   pyridoxal   NH$_2$
                  phosphate

Figure 7.1 Synthesis of GABA.

present in the presynaptic neurone so that increased substrate concentrations do not appear to affect the rate of GABA synthesis. It is possible that the binding of pyridoxal phosphate to GAD is a regulatory control on enzyme activity.

(b) Inhibitors of GAD

These are only of experimental value; they include the pyridoxal antagonists such as **isoniazid** and **methoxypyridoxine, 3-mercapto-propionic acid** (a competitive inhibitor), **allylglycine** which is metabolized to **2,4-ketopentenoic acid**, and **hydroxylamine**. High doses of these compounds also inhibit GABA-transaminase (section 7.2.5(c)). The pyridoxal phosphate antagonists are obviously not specific for GAD but will inhibit all enzymes using the vitamin B$_6$ cofactor.

GABA is also closely related to the oxidative metabolism of carbohydrates in the CNS. In addition to the normal route of glycolysis and oxidation of pyruvic acid by the Krebs (tricarboxylic acid) cycle, the brain can utilize an alternative pathway known as the GABA shunt, the pathways of which are shown in Fig. 7.2. The significance of the shunt either to GABA formation or to energy production is not understood.

7.2.2 Storage

There is at present no direct evidence to indicate that GABA is stored within vesicles in the presynaptic nerve terminal. However, subcellular distribution studies show that most of the GABA is associated with synaptosomes isolated from those regions of the brain which are rich in GABA.

7.2.3 Release

Release of GABA has been demonstrated following electrical stimulation of tissue slices prepared from various regions of mammalian CNS and from the intact brain of anaesthetized animals. Release of GABA is

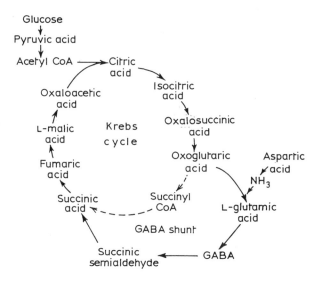

Figure 7.2 Pathways of the GABA shunt. Relationship of GABA to the metabolism of glucose and tricarboxylic acid (Krebs) cycle.

calcium dependent and can be blocked by **tetanus toxin**. Experimentally GABA release from *in vitro* preparations can be inhibited by the presence of GABA agonists which in turn can be blocked by GABA antagonist drugs. This suggests that regulatory GABA autoreceptors may be present on the presynaptic membrane to modulate GABA release.

7.2.4 GABA receptors

Electrophysiological and radioligand binding studies have sub-divided CNS GABA receptors into two subgroups, $GABA_A$ and $GABA_B$. Stimulation of the $GABA_A$ receptors leads to an increase in chloride ion permeability and the induction of hyperpolarization of the postsynaptic membrane. This receptor sub-type, associated with fast inhibitory postsynaptic potentials (IPSPs), is recognized in cerebral cortex and hippocampus. Additionally $GABA_A$ sites are believed to occur pre-synaptically at both axo-axonal and nerve terminal locations, and would be the mechanism by which GABA mediates presynaptic inhibition (section 7.2.4(a)). The $GABA_A$ receptor is stimulated by a number of GABA-like compounds including **muscimol,** but importantly the site is sensitive to competitive antagonism by **bicuculline.**

The $GABA_B$ receptor site is sensitive to the agonist **baclofen** (β-p-chlorophenyl-GABA) but insensitive to the blocking actions of

bicuculline. Stimulation of these receptors likewise mediates hyperpolarization, but this is manifest by changes in potassium conductance and is a slower process than for $GABA_A$ sites. The $GABA_B$ site shows no association with chloride channels. $GABA_B$ receptors may be located on presynaptic terminals where calcium influx into the terminal is reduced which subsequently decreases the synaptic release of transmitter in response to presynaptic nerve stimulation. $GABA_B$ sites have been identified in cerebellum and spinal cord (particularly laminae II and III of dorsal horn).

GABAmodulin is a protein located in the postsynaptic membrane whose function is believed to be analogous to that of the GTP regulatory protein associated with receptors linked to adenylyl cyclase (section 1.9). GABAmodulin activity is determined by its degree of phosphorylation, and this in turn can produce changes in the affinity of GABA receptors for GABA. The benzodiazepine binding site for benzodiazepines and possibly also the barbiturate binding site may also serve to modulate GABA-modulin activity and the GABA receptor complex allosterically.

(a) GABA agonists

The $GABA_A$ receptor is stimulated by a number of experimental GABA-like compounds including **muscimol** (3-hydroxy-5-amino-methyl-isoxazole) and **isoguvacine** P4S and THIP. Muscimol is found in certain fungi including *Amanita muscaria* which contains muscarine; both contribute to toxicity.

The $GABA_B$ receptor is stimulated by **baclofen**, whereas muscimol is only weakly active. GABA receptor agonists are summarized in Table 7.1.

(b) GABA receptor blockers

GABA-receptor blocking agents include **bicuculline** and **picrotoxin,** but these compounds are currently of experimental value only. **Phadophen** and saclophen are weak competitive blockers of $GABA_B$ receptors.

(c) GABA receptors in the CNS

GABA has a widespread distribution in the mammalian CNS, with high concentrations in the hypothalamus, hippocampus and basal ganglia of the brain, in the substantia gelatinosa of the dorsal horn of the spinal cord and in the retina.

Spinal cord. GABA is unevenly distributed in the spinal cord, most being found in the grey matter of the dorsal horn. White matter and nerve

Table 7.1 Cerebral GABA receptor sites

	$GABA_A$	$GABA_B$
Major anatomical location	Cerebral cortex Hippocampus Basal ganglia	Cerebellum Spinal cord
Electrophysiological response	Increased Cl^- conductance Hyperpolarization to fast IPSPs	Increased K^+ conductance Hyperpolarization to slow IPSPs Reduced Ca^{2+} conductance Inhibits adenylyl cyclase
Neuronal site	Axo-axonal junctions Post synaptic	Presynaptic terminals
Agonists	**Muscimol** **GABA** **Isoguvacine** **THIP**	**Baclofen** **GABA**
Specific antagonist	**Bicuculline**	**Phaclofen** **Saclophen**
Associated structures	**Benzodiazepine** binding site **Barbiturate** binding site	

THIP, 4,5,6,7-tetrahydroxyisoxazolo[4,5-c]pyridin-3-ol.

roots contain little GABA. Within the dorsal horn and dorsal column nuclei there is evidence to suggest that GABA is the neurotransmitter of small neurones involved in regulating the activity of primary afferent fibres. GABA neurones are believed to form an axo-axonal synapse onto primary afferent fibres within this region. Release of GABA causes an increase in membrane conductance (i.e. depolarization) in the primary afferent fibre, with the result that the amplitude of action potentials travelling along the primary afferent fibre is greatly diminished, and release of neurotransmitter from primary afferent nerve terminals is reduced (Fig. 7.3). This phenomenon is known as primary afferent depolarization (PAD). The reduction of excitatory transmission in primary afferent fibres by the action of an inhibitory neurotransmitter (GABA) is known as presynaptic inhibition. Presynaptic inhibitory mechanisms may be a major functional role for GABA in the control of spinal reflexes. If GABA function is impaired, central presynaptic inhibitory mechanisms will be decreased, and this may be an important consideration in convulsive, tetanic and spastic disorders.

Cerebellum. The Purkinje cell layer within the cerebellar cortex contains high concentrations of GABA, probably corresponding to the terminals of the Golgi and basket cells synapsing on to the granule and Purkinje cells, respectively. It is believed that the Purkinje cells provide

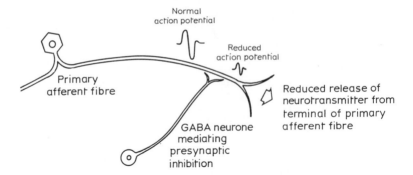

Figure 7.3 Primary afferent depolarization arises because GABA, acting at GABA$_A$ receptors, causes an increase in the conductance of the primary afferent axon by opening Cl$^-$ channels (Cl$^-$ leaves the axon resulting in a net depolarization), and causing reduced Ca^{2+} conductivity at the axon terminal. The increased conductance results in a smaller action potential arriving at the nerve terminal. This, and the reduced Ca^{2+} conductivity, results in decreased transmitter release from the primary afferent terminal, and is responsible for the observed presynaptic inhibition (of release of transmitter from the primary afferent terminal) by GABA.

a GABA-ergic inhibitory outflow to Deiters' nucleus (lateral vestibular nucleus) in the medulla, a site also having a high GABA concentration. Such cerebellar locations of GABA indicate a possible functional role for this neurotransmitter in the control of cerebellar reflexes and various inhibitory mechanisms on spinal alpha motoneurones.

Basal ganglia. High concentrations of GABA are found in the basal ganglia, particularly in the substantia nigra, globus pallidus and nucleus accumbens. Most GABA in the caudate nucleus and putamen is associated with short inhibitory interneurones, while a long-axon GABA-mediated pathway projects from the globus pallidus and striatal complex to the substantia nigra. This GABA-mediated pathway is believed to exert an inhibitory feedback controlling influence on the ascending dopaminergic nigro-striatal projection, and thus to play an important role in extrapyramidal function (see Figure 4.4).

Retina. The horizontal cell layer in the retina is rich in GABA, and in some species this amino acid has been proposed as the mediator of feedback and lateral inhibition in the retina produced by the horizontal cells.

Thalamus. Many thalamic nuclei contain high concentrations of GABA, and GABA may be involved in control of sensory traffic through the thalamus.

(d) Benzodiazepines and GABA_A receptors

Biochemical and electrophysiological studies indicate a close relationship of benzodiazepines (and barbiturates) and GABA$_A$ receptors. Since their introduction into clinical use, benzodiazepines have been used as anxiolytics, hypnotics, muscle relaxants, anticonvulsants and sedatives. The choice of agent for a specific application has in the past been somewhat empirical, because at the right dose, most benzodiazepines will exhibit the full spectrum of properties. A major consideration with the clinical use of benzodiazepines is the formulation, plasma half-life of the compound, and mechanism of elimination if this leads to formation of active metabolites.

Most benzodiazepines are not water soluble; this presents no problem if they are taken orally as lipid solubility is more important when drugs are taken by this route. For rapid onset of action, **diazepam** is available as an emulsion (**diazemules**) which is given intravenously for rapid control of status epilepticus; **midazolam** is a water soluble benzodiazepine suitable for intravenous injection.

Benzodiazepines are inactivated mainly by metabolism by microsomal enzymes. The duration of action is related to rate of metabolism to inactive metabolite.

Nitrazepam has a plasma half-life of about 24 h; it does not form active metabolites but has nonetheless a long duration of action which is not compatible with its promotion as an hypnotic.

Temazepam and **triazolam** are marketed as hypnotics; they have plasma half-lives of about 8 h and do not form active metabolites, so might be expected to have a lower incidence of hangover effects after use the night before. For similar reasons, the duration of drowsiness will be shorter if these compounds are used (in place of more lasting compounds), when only a short period of sedation is required (e.g. for dental surgery).

Diazepam has been the most frequently used benzodiazepine, although its use for short procedures seems unjustifiable, as it has a long plasma half-life (1–2 days) and is metabolized to **desmethyldiazepam** which in turn is converted to **oxazepam**, both of which are active metabolites with plasma half-lives of 2–4 days and 12 h respectively.

The potency of benzodiazepines both clinically and experimentally would predict specific and selective action in the CNS, and indeed the early studies showed that they effectively enhance the electrophysiological and biochemical actions of GABA. Radiolabelled ligand binding studies confirmed an increase in specific GABA binding to GABA receptors in the presence of benzodiazepines. Further, the **benzodiazepines** were shown to be concentrated in brain regions rich in GABA (cortex, cerebellum and limbic structures including hippocampus and amygdala), and hence a specific benzodiazepine binding site was proposed. More recent information has led to the hypothesis of a functional membrane-bound GABA receptor complex in which the GABA binding site is in close association with a benzodiazepine recognition site and the Cl^- channel. This macromolecular complex may also contain a barbiturate and picrotoxin binding site. Many researchers would now suggest that the benzodiazepines do not have direct actions of their own but serve to modify the action of GABA on the permeability of the chloride ion channel. Since the identification of a benzodiazepine recognition site, researchers have sought to isolate the endogenous receptor ligand; current suggestions include the family of β-**carbolines** and **desmethyldiazepam** which has been shown to occur naturally in brain tissue, while the peptide Diazepam Binding Inhibitor (DBI) has been shown to interfere with **benzodiazepine** binding and subsequent GABA action.

There is some debate as to whether two (or more) types of **benzodiazepine** receptors exist. On the basis of differential agonist and antagonist binding patterns a BZ type 1 and BZ type 2 receptor have been proposed although their functional significance (if any) remains undefined.

Although some benzodiazepine antagonist compounds exist experimentally only one drug is currently available clinically. **Flumazenil** is a newly-developed short-acting benzodiazepine antagonist which is marketed for the diagnosis, the reversal and treatment of benzodiazepine overdose where unconsciousness or respiratory depression are a danger to the patient. By itself, **flumazenil** has no apparant clinical action.

7.2.5 Inactivation of released GABA

(a) Uptake of GABA

GABA is removed from the extracellular space by a high-affinity, sodium-dependent uptake system present in both presynaptic GABA nerve terminals and surrounding glial elements.

(b) Inhibitors of GABA uptake

Many experimental compounds inhibit GABA uptake; some preferentially inhibit the neuronal uptake of GABA (e.g. **diaminobutyric acid, cis-1,3-aminocyclohexane carboxylic acid**), while others block glial uptake systems (e.g. β-**alanine**). Other compounds are less selective and inhibit both neuronal and glial uptake of GABA (e.g. **nipecotic acid**). In addition, the benzodiazepines, neuroleptics and tricyclic antidepressants inhibit GABA-uptake mechanisms, although the significance of this action with respect to their clinical effectiveness is not known.

(c) Enzymatic metabolism of GABA

Presumably much of the GABA taken back into the presynaptic neurone following release and receptor interaction is recycled as releasable transmitter. In addition, GABA can be enzymatically catabolized in both the nerve terminal and glial tissue. GABA is first converted in the presence of α-oxoglutaric acid into succinic semialdehyde by the mitochondrial enzyme GABA-aminotransferase (GABA-α-oxoglutarate transminase, GABA-T). GABA-T requires pyridoxal phosphate as cofactor. Succinic semialdehyde is rapidly oxidized to succinic acid by the enzyme succinic semialdehyde dehydrogenase, in a reaction which requires $NAD^+/NADH$ as cofactor (Fig. 7.4). The succinic acid so formed enters the tricarboxylic acid (Krebs) cycle and intermediary metabolism.

(d) Inhibitors of GABA-T

There are a number of experimentally-available compounds which inhibit GABA-T (e.g. **ethanolamine-O-sulphate, γ-acetylenic GABA, γ-vinyl GABA, gabaculine, hydrazinopropionic acid**), and the treatment of animals with such compounds results in gross increase in cerebral GABA concentrations. Until recently the only putative inhibitor of GABA metabolism in clinical use was **sodium di-*n*-propylacetate (sodium valproate)**, which seems effective in some patients with epileptic or psychomotor disorders, although it is not certain whether its therapeutic actions are solely due to effects on the GABA system. As GABA-T requires vitamin B_6 as cofactor, the enzyme is also inhibited by pyridoxal phosphate inhibitors (e.g. **amino-oxyacetic acid**).

Recently, **vigabatrin (γ-vinyl-GABA)**, an irreversible inhibitor of GABA-T has been introduced for treatment of epilepsy. Its efficacy in treatment of epilepsy will become apparent with more extensive clinical use.

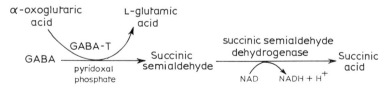

Figure 7.4 Metabolism of GABA.

7.2.6 GABA and its proposed relation to various clinical states

Experimental observations have led to the suggestion that several clinical conditions may be related to malfunctions of GABA systems. These disorders, together with drugs currently used to alleviate the symptoms, are considered below. It will be seen that future development of GABA-specific drugs could be of vital importance in management of certain CNS disorders, although the reader must be reminded that the associations of GABA-ergic neurotransmission to pathological states remain tenuous.

(a) Extrapyramidal movement disorders

Substantial loss of GAD and GABA in basal ganglia has been measured in post-mortem material from patients dying from Huntington's chorea and more recently, reduction of GABA/benzodiazepine binding sites has been shown. Loss of such GABA-mediated inhibitory mechanisms within the basal ganglia may result in the disinhibited dyskinetic movements associated with this disease. Unfortunately there are no suitable GABA-mimetic drugs (or indeed any other therapeutic agents) currently available for effective treatment of this distressing disorder. Neuroleptics (e.g. **reserpine, tetrabenazine, phenothiazines, butyrophenones** and **pimozide**) and benzodiazepines, are employed for their anti-hyperkinetic and sedative properties (section 4.6).

Changes in GABA biochemistry have also been suggested in the aetiology of Parkinson's disease and the neuroleptic-induced tardive dyskinesias. Some loss of GABA and GAD is reported from the Parkinsonian brain but whether this is a specific effect or related to a general loss of neurones from the basal ganglia is not firmly established.

(b) Disorders of spinal cord function and muscle tone: spasticity

Impairment of GABA neurotransmission within the spinal cord (whether by a local lesion or loss of higher controlling influence as in an upper motor neurone lesion), often results in disruption of presynaptic inhibitory mechanisms with subsequent development of muscle spasticity. **Baclofen** and the benzodiazepines (e.g. **diazepam** and **clonazepam**), which are claimed to possess some GABA-mimetic properties, are used for their antispastic and muscle-relaxant effects. These beneficial actions are believed to arise through an enhancement of the presynaptic inhibition of the inputs to spinal motoneurones. **Diazepam** also relieves athetosis in patients with cerebral palsy, and can be effective in the treatment of torticollis.

(c) Epilepsy

Administration of drugs which block central GABA mechanisms (e.g. GAD inhibitors, GABA-receptor blocking agents) induce convulsions in animals, while drugs enhancing GABA neurotransmission (e.g. GABA-T inhibitors, benzodiazepines, barbiturates) usually decrease the susceptibility of animals to seizures. These observations have been the basis for a possible role of GABA in the aetiology of various forms of epilepsy.

Drugs used in the management of epilepsy are called anticonvulsant (anti-epileptic) drugs. **Vigabatrin** is the only real GABA-mimetic drug in clinical use, and its usefulness is not yet established. Experimentally, the barbiturates (e.g. **phenobarbitone, primidone**) have been shown to enhance GABA-ergic mechanisms in the CNS, and this in part may explain their clinical properties. **Sodium valproate** is effective in the management of most forms of epilepsy and, although experimentally it can inhibit the activity of GABA-T and succinic semialdehyde dehydrogenase, it is not certain how much this action contributes to its anticonvulsant effects. The usefulness of the benzodiazepines (**clonazepam, clobazam, diazepam**) as anticonvulsants is presumably related to their GABA-enhancing actions in the CNS. Other well-established and commonly used anticonvulsant drugs (e.g. **phenytoin, carbamazapine**) depress the activity of most excitable membranes by blocking voltage regulated Na^+ channels.

(d) Degenerative states: pre-senile (Alzheimer's disease) and senile dementias

The dementias are characterized clinically by memory impairment and varying degrees of confusion. Pathologically they relate to loss of cortical neurones and biochemical studies confirm low concentrations of GABA in the demented brain in keeping with loss of interneurones. It is not known whether loss of this neurotransmitter is related primarily or secondarily to the aetiological process. This finding has certainly stimulated much drug research in attempts to develop possible GABA-enhancing agents which may be of value in the management of this very disabling complaint.

(e) Thought and emotional disorders

GABA mechanisms have been implicated in both schizophrenia and anxiety states. Changes in GABA biochemistry have been reported in both animal models and post-mortem material of schizophrenia.

However it is also known that long-term neuroleptic therapy can modify GABA neurotransmission. Likewise the anxiolytic effects of the benzodiazepines is clearly well established, and there are a number of theories to link changes in GABA receptor mechanisms with the aetiology of anxiety.

Tetanus toxin can cause prolonged convulsion by an action within the spinal cord, where it is known to decrease the release of inhibitory amino acid transmitters (GABA and glycine). Centrally acting muscle relaxants, anticonvulsants and tubocurarine-like muscle relaxants are used to treat this condition.

7.3 GLYCINE

Glycine is believed to be an important inhibitory neurotransmitter in the mammalian spinal cord. Its possible role at retinal and supraspinal sites has yet to be firmly established.

7.3.1 Synthesis

Glycine is found in the free form in all tissues of the body. It is an important constituent of the diet, but is also available from the breakdown of peptides and proteins and of nucleotides and nucleic acids. Glycine can be formed from carbohydrates by way of 3-phosphoserine and serine, and it is readily interconvertible with serine in the CNS by the enzyme serine hydroxymethyl transferase (SHMT), which utilizes tetrahydrofolic acid as cofactor (Fig. 7.5).

7.3.2 Storage

The possible mechanisms of the storage of neurotransmitter glycine within presynaptic nerve terminals have not been established.

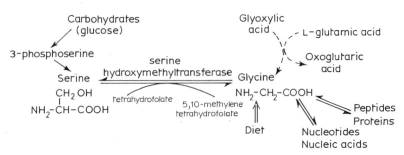

Figure 7.5 Synthesis and sources of glycine.

7.3.3 Release

Release of glycine has been demonstrated from slices of spinal cord following either electrical stimulation or exposure to high external potassium concentration. Such release is partially calcium dependent, and is reduced by **tetanus toxin**.

7.3.4 Receptors

Glycine decreases the rate of firing of postsynaptic neurones, by causing an increase in membrane conductance to chloride ions.

(a) Glycine receptors in the CNS

The highest concentrations of glycine found within the CNS are located in the spinal cord and brainstem. In the spinal cord higher concentrations are found in grey matter than in white matter. Electrophysiologically, glycine inhibits certain neurones in the spinal cord dorsal column nuclei, brainstem (medullary reticular formation), retina and diencephalon (forebrain). Accordingly, the consensus is that glycine is the neurotransmitter of certain small inhibitory interneurones in the brainstem and spinal cord (and possibly at some sites supraspinally where its functional role has still to be determined). The effects of glycine in the spinal cord are always believed to be postsynaptic and therefore, unlike GABA, glycine does not mediate presynaptic inhibition.

Glycine is believed to be the neurotransmitter of the inhibitory Renshaw cell in the ventral horn of the spinal cord. The Renshaw cell is

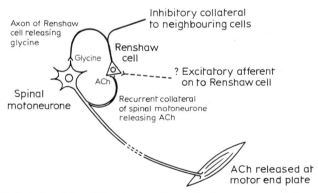

Figure 7.6 Relationship of Renshaw cell to spinal motoneurone in ventral horn of the spinal cord.

believed to receive an excitatory input from recurrent collaterals of spinal motoneurones, an event mediated by acetylcholine release (Fig. 7.6). Axons from the Renshaw cell in turn synapse on the motoneurones. Thus, glycine appears to be the neurotransmitter of recurrent inhibition of motoneurones in the mammalian spinal cord. At these and other interneuronal synapses, glycine is possibly involved in the regulation of spinal and brainstem reflexes. Clinically, glycine may have a role in motoneurone spasticity, since experimental occlusion of the blood supply to the spinal cord in animals results in spastic paraplegia together with marked falls in spinal glycine concentrations; spinal GABA levels are not affected.

(b) Glycine-receptor blocker

The inhibitory actions of glycine are competitively blocked by **strychnine**. Administration of **strychnine** in small doses to man or animals causes enhanced motor responses of spinal reflexes: larger doses induce convulsions. This convulsant effect of **strychnine** is mediated spinally and is believed to result from blockade of inhibitory glycinergic sites. **Strychnine** also acts in the medulla where it stimulates vasomotor and vagal centres. Death is due to asphyxia resulting from repeated convulsions. **Strychnine** poisoning can often be successfully treated by intravenous administration of an anticonvulsant such as **diazepam**. No agonists at the glycine receptor are known.

7.3.5 Inactivation of released glycine

(a) Uptake of glycine

Glycine is transported into tissues, particularly in the CNS, by two uptake mechanisms: a low-affinity uptake system which is shared with other small neutral amino acids (e.g. serine, leucine, proline, alanine), and a high-affinity, energy-requiring, sodium-dependent system specific for glycine.

(b) Metabolism of glycine

Little is known about the possible enzymatic metabolism of glycine, or the possible redirection of transmitter glycine into other metabolic pools. It can be converted to serine by SHMT (Fig. 7.5).

7.4 TAURINE

Taurine is the most abundant amino acid in the body, although it is not incorporated into proteins. It has tentatively been suggested that it may have a neurotransmitter (or neuromodulatory) role in some parts of the nervous system.

Taurine is a sulphur-containing amino acid which is consumed in the diet but can also be synthesized from cysteine which is oxidized to cysteine sulphinic acid by cysteine oxidase which in turn is decarboxylated to hypotaurine; hypotaurine is finally oxidized to taurine (Fig. 7.7). Some taurine appears to be associated with nerve endings and can be released in response to both electrical and chemical stimulation in a calcium-dependent manner.

Taurine has a depressant action on postsynaptic cells, and is believed to act as a membrane stabilizer linked to the Cl^- channel. This depressant action of taurine can be blocked by **strychnine** (section 7.3.4), but not by GABA-receptor blockers such as **bicuculline**.

A high-affinity sodium-dependent uptake system has been demonstrated for taurine in both neurones and glial elements. The half-life of taurine in the body is very long, indicating that the metabolism of this amino acid is very slow.

7.4.1 Distribution and possible functions of taurine

The brain, retina and striated muscle contain large concentrations of taurine. However, not all organs contain the capacity of being able to synthesize this amino acid although many can accumulate it from the general circulation. The functional role of taurine in the body is at present unknown. It may act as an inhibitory neurotransmitter at certain sites such as spinal cord, brainstem and retina. Other workers prefer to regard taurine as a modulator rather than a true neurotransmitter. High concentrations of taurine have been noted in the plasma and urine of

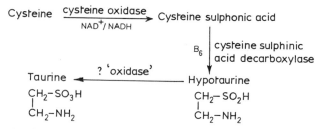

Figure 7.7 Synthesis of taurine from cysteine.

patients suffering from diseases of muscle. The possible role of this amino acid in the muscular dystrophies, however, remains speculative.

Perhaps the most convincing site where taurine may support a neurotransmitter role is in the retina, at inhibitory synapses in the inner plexiform layer. In addition, taurine is found in high concentration in the photoreceptor layer of the retina and, although it is not considered to be a neurotransmitter at this site, feeding animals on diets deficient in taurine leads to the degeneration of photoreceptors in the retina: this effect can be reversed by the addition of taurine-rich food.

FURTHER READING

Bormann, J. (1988) Electrophysiology of GABA$_A$ and GABA$_B$ receptor subtypes, *Trends Neurosci.*, 11, 111–12.

Bowery, N. (1989) GABA$_B$ receptors and their significance in mammalian pharmacology. *TIPS*, 10, No. 10.

Lader, M. (1987) Clinical pharmacology of benzodiazepines. *Ann. Rev. Med.* 38, 19–28.

Lal, H., Fielding, S., Malick, J., Robert, E., Shah, N. and Usdin, E. (eds) (1980) *GABA Neurotransmission: Current Developments in Physiology and Neurochemistry.* Ankho Int., New York.

Meldrum, B. (1987) Classification of GABA and benzodiazepine receptors. *J. Psychopharmacol.,* 1, 1–5.

Reynolds, E.H. (1990) Vigabatrin. *Brit. Med. J.,* 300, 277–8.

Roberts, E., Chase, T.N. and Tower, D.B. (eds) (1976) *GABA in Nervous System Function.* Raven Press, New York.

Sieghart, W. (1989) Multiplicity of GABA$_A$-benzodiazepine receptors. *TIPS*, 10, No. 10.

Tallmann, J.F. and Gallager, D.W. (1985) The GABAergic system: a locus of benzodiazepine action. *Ann. Rev. Neurosci.,* 8, 21–44.

8 Excitatory amino acids: L-glutamic acid and L-aspartic acid

There have recently been major research advances to support the neurotransmitter roles for both L-glutamic acid and L-aspartic acid within the mammalian nervous system. Prior to this development it has long been known that glutamic acid has excitatory electrophysiological actions on central neurones, and its neurotransmitter status in the invertebrate nervous system remains unchallenged. Recent studies have convincingly demonstrated selective actions for these compounds on mammalian cells and this has led to the classification of excitatory amino acid receptor sub-types. The possible functional roles of these receptor sub-groups remain to be fully elucidated, but initial observations have suggested novel associations and potentially close involvement with the aetiology of a number of neurodegenerative disorders.

8.1 SYNTHESIS

Glutamate and aspartate are important constituents of the diet, and both of these dicarboxylic amino acids are associated with a number of metabolic processes within the cell. L-Glutamic acid is the most abundant amino acid in the CNS and can be synthesized from many different sources including:

1. From α-oxoglutarate and aspartate via the enzyme aspartate aminotransferase;
2. From glutamine by phosphate-activated glutaminase;

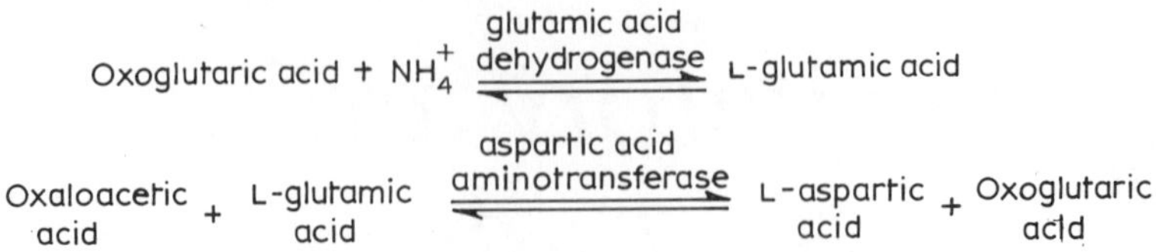

$$\text{Oxoglutaric acid} + NH_4^+ \xrightarrow{\text{glutamic acid dehydrogenase}} \text{L-glutamic acid}$$

$$\underset{\text{acid}}{\text{Oxaloacetic}} + \underset{\text{acid}}{\text{L-glutamic}} \xrightarrow{\text{aspartic acid aminotransferase}} \underset{\text{acid}}{\text{L-aspartic}} + \underset{\text{acid}}{\text{Oxoglutaric}}$$

Figure 8.1 Synthesis and metabolism of glutamic acid.

3. From reaction of α-oxoglutarate with NH_4^+ catalysed by glutamic acid dehydrogenase (Fig. 8.1).

L-Aspartic acid can be formed by the transamination of oxaloacetic acid and L-glutamic acid by the enzyme aspartate aminotransaminase (Fig. 8.1).

Biochemical studies suggest that several metabolic 'pools' of excitatory amino acids exist within the CNS but that none appears specifically to be associated with a neurotransmitter role.

8.2 STORAGE

The storage of glutamate and aspartate is ill understood, but immuno-chemical techniques have demonstrated the probable presence of glutamate-like compound within dense-core vesicles associated with nerve terminals.

8.3 RELEASE

Release of L-glutamic acid and L-aspartic acid has been demonstrated from slices and synaptosomes of various brain regions (cerebral cortex, hypothalamus) and spinal cord in response to either electrical stimulation or raised extracellular potassium concentration. Such release of these excitatory amino acids has been shown to be calcium dependent.

8.4 EXCITANT AMINO ACID RECEPTORS IN THE CNS

L-Aspartate and L-glutamate occur throughout the CNS, although both amino acids demonstrate an uneven distribution within the brain and spinal cord. Lesion studies suggest a neurotransmitter role for one or both of these amino acids in excitatory pathways in most regions of the CNS. For example in the spinal cord L-glutamate is most concentrated at primary afferent fibres in the dorsal roots, and it is believed the amino acid may serve to relay sensory information and to regulate motor activity and spinal reflexes at these sites. In the brain L-glutamic acid is found in high concentrations in the cerebral cortex, hippocampus, neostriatum and cerebellum, with lower levels in the hypothalamus. In the dorsal and ventral grey matter L-aspartate has been proposed as an excitatory neurotransmitter at spinal excitatory interneurones where it may regulate motor and spinal reflexes.

Intracerebral administration of L-aspartic acid or L-glutamic acid to animals induces convulsions, although positive links between these excitatory amino acids and epilepsy remain to be firmly established. L-aspartate and L-glutamate are also found in high concentrations in the mammalian retina, where they may function as excitatory neurotransmitters at photoreceptors.

There is no good evidence for the association of the 'Chinese Restaurant Syndrome', and glutaminergic neurotransmission. The symptoms of the syndrome (tightness of the chest, tachycardia, lacrimation, gastro-intestinal discomfort and nausea) cannot be replicated under laboratory conditions when large quantities of monosodium glutamate (a flavouring agent much used in Western as well as Chinese cuisine) are eaten. The symptoms are not all that dissimilar to those which occur following ingestion of immoderate quantities of ethanol in a hot and stuffy atmosphere (under which conditions the symptoms can all too readily be replicated).

8.4.1 Excitatory amino acid receptor agonists and antagonists

Biochemical and electrophysiological studies, using agonists, divide excitatory amino acid receptors into three sub-types, namely, N-methyl-D-aspartate (NMDA) receptor, the quisqualate receptor, and the kainate receptor.

Selective agonists for the NMDA receptor include **N-methyl-D-aspartate, aspartate** and **ibotenate**: indeed NMDA is some 1000 times more potent than glutamate at this site. Agonists at the quisqualate receptor include **quisqualate, glutamate** and **AMPA** (α-amino-3-hydroxy-5-methyl-isoxazole propionic acid); agonists for the kainate receptor include **kainate, domoate** and **glutamate** (Table 8.1).

Selective antagonists at the NMDA receptor have been described and include **D-AP5** and **CPP** (D-2-amino-5-phosphonopentanoic acid and 3-3(2-carboxypiperazine-4-yl-propyl-1-phosphonic acid) respectively). Compounds interacting at excitatory amino acid receptor are summarized in Table 8.1.

The best characterized excitatory amino acid receptor sub-type is the NMDA receptor, so much so that the other two sub-groups are often

Table 8.1 Cerebral excitatory amino acid receptor subtypes

Receptor type	NMDA receptor	Non-NMDA receptor	
	NMDA	Quisqualate	Kainate
Main anatomical location	Hippocampus Basal ganglia Limbic system Cerebral cortex (outer layers) Superior colliculus Vestibular nuclei Cerebellum	Hippocampus Basal ganglia Limbic system Cerebral cortex (outer layers)	Hippocampus Basal ganglia Thalamus/hypothalamus Cerebral cortex (inner layers)
Agonists	NMDA Aspartate Ibotenate Glutamate	Quisqualate AMPA Glutamate	Kainate Domoate Glutamate
Antagonists	D-AP5 CPP	DNQX CNQX	DNQX CNQX
Electrophysiological response	Influx of Na^+, K^+, Ca^{2+} slow prolonged depolarization	Flux of Na^+, K^+	Flux of Na^+, K^+, Ca^{2+} fast depolarization
Associated structures	Glycine binding site		
Ketamine MK801 } Phencyclidine	NMDA open channel blockers		

CPP: 3-(2-carboxypiperazine-4-yl)-propyl-1-phosphonic acid.
AP5: 2-amino-5-phosphonopentanoic acid.
AMPA: α-amino-3-hydroxy-5-methylisoxazole-propionate.
CNQX: 6-cyano-7-nitroquinoxaline-2,3-dione (FG9065).
DNQX: 6,7-dinitroquinoxaline-2,3-dione (FG9041).

classed together as the non-NMDA (quisqualate and kainate) receptor. Whereas NMDA and non-NMDA sites are often found in similar anatomical locations (Table 8.1), they differ electrophysiologically. The non-NMDA receptor is associated with fast depolarization of the cell membrane related to an influx of Na^+ ions. The NMDA receptor on the other hand causes a slower, more prolonged depolarization due to movement of Na^+, K^+ and Ca^{2+} ions across the membrane, a process which appears to be regulated by Mg^{2+} ions in a voltage-dependent manner. Some NMDA receptors are believed to be linked to other neurotransmitter binding sites, e.g. glycine, and this hypothesis is discussed below (Fig. 8.2). The quisqualate and kainate receptors appear to exist

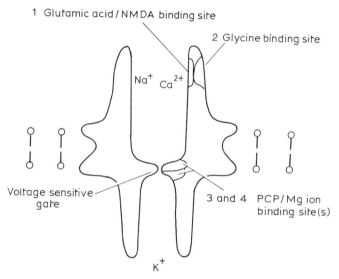

Figure 8.2 The NMDA receptor–ion channel complex. Electrophysiological and radioligand binding studies have helped in the elucidation of the proposed structure of the NMDA receptor–ion channel complex. Four main sites within this receptor complex are recognized:

1. The transmitter recognition site to which agonist and antagonist drugs bind. The interaction of agonists at this site is associated with activation of the receptor complex and opening of the ion channel.
2. A glycine binding site on the outer aspect of the NMDA receptor which appears to allosterically potentiate the action of agonists at the transmitter recognition site and thereby enhances ion channel opening.
3. A cation binding site located inside the ion channel to which Mg^{2+} ions can bind to block transmembrane ion fluxes.
4. A phencyclidine (PCP) binding site again within the ion channel to which a number of drugs such as σ-opiates and dissociative anaesthetics (e.g. PCP, ketamine) can bind to block receptor-ion channel function.

in two states, high-affinity and low-affinity, which show differential binding affinities for the excitatory amino acid agonists.

8.4.2 Kainic acid and excitotoxicity

Kainic acid has a neurotoxic action on certain neurones in the CNS. It is believed to be excitotoxic by way of its potent agonist actions at certain excitatory amino acid receptors. Following injections of high concentrations of this drug locally into the brain, the neurotoxic action of kainic acid is apparently specific to neuronal cell bodies (which possess excitatory amino acid receptors) causing these and their projecting axons to degenerate (see also section 8.4.3 on long-term potentiation). Nerve axons and terminals adjacent to the injection site which do not carry kainate binding sites are left intact. Investigation of the mechanism of kainate-mediated excitotoxicity highlights the movement of calcium ions into the cell with an initial acute Ca^{2+}-independent process associated with influx of Na^+ ions and fast depolarization followed by the neurotoxic Ca^{2+}-dependent sequence.

Excess intracellular concentrations of Ca^{2+} lead to unregulated stimulation of a wide range of calcium-dependent mechanisms. The metabolic load imposed on the cell cannot be sustained and cell death results. Excitotoxicity is characterized by frequent neuronal depolarization, epileptiform spiking and an unsustainable increase in oxygen demand.

Excitotoxicity occurs following damage to the CNS, however caused; mechanical, chemical, ischaemic, infarct (stroke), infection or injury can lead to neuronal death following excitotoxicity. A common consequence of damage to the CNS is release of glutamic acid in the area which has sustained damage.

Excitotoxicity may be a feature of some forms of epilepsy; competitive and non-competitive inhibitors of excitatory amino acid receptors are generally active in all animal models of epilepsy. One problem with most antagonists of excitatory amino acids is that they are not sufficiently lipid soluble to allow adequate concentration in the CNS in man (in animal experiments; they can be administered intracerebrally).

8.4.3 Long-term potentiation (LTP)

As the name implies, LTP is the long-term enhancement of the synaptic response to presynaptic neuronal stimulation, believed to be related to the plasticity of synapses for excitatory amino acids. Under the right conditions, LTP can last several hours, or even days or weeks. It is proposed that certain NMDA receptors may be involved in the remarkable electrophysiological and behavioural results due to LTP in hippocampal neurones. Certainly impairment of NMDA-linked ion channel function with the dissociative anaesthetics blocks the induction of LTP, as do selective NMDA antagonists.

Mechanism of LTP

In hippocampal neurones, low frequency stimulation causes the release of L-glutamate from presynaptic nerve endings and evokes a normal EPSP believed in general to be mediated by activation of non-NMDA receptors. (Under normal conditions, the NMDA receptor channel complex is blocked by the presence of Mg^{2+} ions in the synaptic cleft.) However with high-frequency stimulation, there is widespread dendritic depolarization with release of sufficient glutamate to overcome the Mg^{2+}-induced blockade of the NMDA channel: Ca^{2+} now enters the cell through the ion channel and activates the intracellular biochemical processes involved in LTP. Thus an increase in neurotransmitter release (or experimentally a depletion of synaptic Mg^{2+}) appears to be involved in the initiation and maintenance of LTP with the subsequent widespread activation and disinhibition of NMDA sites. LTP can be blocked with NMDA receptor blockers such as AP5.

8.5 INACTIVATION OF RELEASED EXCITATORY AMINO ACIDS

8.5.1 Uptake of aspartate and glutamate

L-Aspartic acid and L-glutamic acid are taken up into tissues by two uptake processes, a low-affinity transport system and a high-affinity, sodium-dependent uptake system. It is the high-affinity system which is believed to be located specifically at the terminals of the excitatory neurones, although the role of glial tissue in the removal of the amino acids from the synaptic cleft must be further investigated. Both low and high-affinity uptake systems can be blocked by high concentrations of chlorpromazine.

8.5.2 Metabolism of aspartate and glutamate

One metabolic pathway for L-glutamic acid involves its decarboxylation to GABA (section 7.2.1). In another reaction it may undergo transamination with oxaloacetic acid to form α-oxoglutaric acid and L-aspartic acid. A further aspect of the metabolism of L-glutamic acid is its conversion into glutamine by the enzyme glutamine synthetase (Fig.

$$\text{L-glutamic acid} + NH_3 \xrightarrow{\text{glutamine sythetase}} \text{glutamine}$$

Figure 8.3 Metabolism of L-glutamate.

reclaiming and storing synaptically released L-glutamic acid. Excess L-aspartic acid is probably metabolised to various intermediates of the Krebs (tricarboxylic acid) cycle.

8.6 FUNCTIONAL ROLES OF CEREBRAL EXCITATORY AMINO ACID NEUROTRANSMITTERS

With the demonstration of excitatory amino acid receptor sub-groups and heterogeneity in anatomical distribution within the mammalian brain, a number of functional roles have been postulated for these neurotransmitters. However, as yet the precise roles are unknown but some recent and exciting concepts are discussed below.

8.6.1 Neurodegenerative disorders

The excitotoxin action of kainic acid on cells is well described (section 8.4.2). Extrapolation of this concept has led to the hypothesis that high, local concentrations of the excitatory amino acid neurotransmitters within the brain and spinal cord might, through excitotoxic mechanisms, play a role in the aetiology of human neurodegenerative disease including Alzheimer's disease and the other dementias, Huntington's chorea, olivopontino-cerebellar degeneration, and brain damage associated with anoxia/ischaemia, hypoglycaemia, stroke and epilepsy. Acute neuronal dysfunction may be related to release of excessive quantities of excitatory amino acids onto vulnerable neurones to act as endogenous excitotoxins to cause neuronal destruction. Although the type of excitatory amino acid receptor involved in this neurodegenerative process remains speculative, it leads to the suggestion that early intervention with an appropriate excitatory amino acid receptor antagonist may arrest and thereby protect against these neurodegenerative processes. Indeed initial limited clinical trials have begun to investigate the possible effectiveness of proposed receptor blocking compounds in stroke. Certainly in animal models of cerebral ischaemia, selective NMDA receptor antagonists significantly reduce the resultant neuronal damage secondary to anoxia. It is still too early to deliberate on the hoped success of such drugs in limiting the size of cerebral infarction and thereby of the neurological deficits in stroke, but scientists and clinicians alike are optimistic. To pursue this exciting concept further we must await the development of lipid soluble, specific and selective antagonists at excitatory amino acid receptors.

8.6.2 Learning and memory

The hippocampus is believed to be a major centre for processing memory and for cognitive function. It is proposed that many of its neuronal systems use the excitatory amino acids, and therefore these neuro-transmitters presumably function in the role in learning and memory. This concept is supported from findings that changes in kainate receptors (losses from some layers of hippocampus and increases in other layers) are noted in Alzheimer's disease. Further, hippocampal NMDA receptors appear to be important in learning behaviour in animals, the selective NMDA antagonist AP5 blocking both long-term potentiation (LTP) and learned patterns, thereby causing amnesia.

In degenerative states such as Alzheimer's disease, plastic changes appear to occur in response to pathological damage, and compensatory mechanisms such as dendritic sprouting seem to operate. Hence both increases and decreases in excitatory amino acid receptors have been measured in this disorder. It is not known whether the pathological damage in dementia is related to excitotoxic-mediated cell death, but it does raise the issue as to whether treatment with appropriate excitatory amino acid receptor blockers could arrest the process of degeneration.

8.6.3 Other neurological and psychiatric states

The widespread distribution of the excitatory amino acid receptors within the CNS has led to the inevitable proposal that these potential neurotransmitters are involved in the aetiology of many neurological and psychiatric conditions. From their location in limbic areas, the involvement of glutamate and aspartate in schizophrenia and other emotional disorders is implicated. The control of spasticity and pain pathways is suggested by the demonstration of large numbers of excitatory amino acid receptors in the spinal cord. However, the true functional roles of the amino acids at these sites have yet to be fully elucidated.

FURTHER READING

Collingridge, G.L. and Bliss, T.V.P. (1987) NMDA receptors – their role in long-term potentiation. *Trends Neurosci.*, **10**, 288–90.
Deary, I.J. and Whalley, L.J. (1988) Recent research on the causes of Alzheimer's disease. *Brit. Med. J.*, **297**, 807–10.

Foster, A.C. (1988) Quisqualate receptor antagonists. *Nature*, **335**, No. 6192.

Foster, A.C. and Fagg, G.E. (1987) Taking apart NMDA receptors. *Nature*, **329**, 395–6.

Lodge, D. (1988) *Excitatory Amino Acids in Health and Disease*. Interscience Publication, John Wiley, New York.

Maycox, P.R., Hell, J.W. and Jahn, R. (1990) Amino acid neurotransmission: spotlight on synaptic vesicles. *TINS*, **13**, No. 3.

Monaghan, D.T., Bridges, R.J. and Cotman, C.W. (1989) The excitatory amino acid receptors. *Ann. Rev. Pharmacol.*, **29**, 365–402.

Rothman, S.M. and Olney, J.W. (1987) Excitotoxicity and the NMDA receptors. *Trends Neurosci.*, **10**, 299–302.

Watkins, J.C. and Evans, R.H. (1981) Excitatory amino acid transmitters. *Ann. Rev. Pharmacol.*, **21**, 165–204.

Watkins, J.C. and Olverman, H.J. (1987) Agonists and antagonists for excitatory amino acid receptors. *Trends Neurosci.*, **10**, 265–72.

9 Peptides and neuronal function

9.1 PEPTIDES AS NEUROTRANSMITTERS AND HORMONES

In 1936, von Euler described the properties of an extract prepared from brain and gut which had the form of a white powder. This extract became known as substance P (SP), and it is discussed later in this chapter. The observation which was to prove to be of great significance was that the biological activity of substance P could be destroyed by trypsin, thus showing that substance P was a peptide. Some 30 years later, Harris showed that the release of hormones from the anterior pituitary was under the control of blood-borne factors. These blood-borne factors were released from the hypothalamus and carried to the anterior pituitary by a portal system. These factors were shown to be released from neurones in the hypothalamus, and furthermore, they were shown to be peptides. These observations led to the suggestion that peptides released from nerves could act as neurotransmitters.

With the development and application of more advanced bioassay, immunological, histochemical and other modern techniques in protein biochemistry and molecular biology (in particular with the development of DNA gene probes and the use of monoclonal antibodies to raise anti-serum to small peptide molecules), there has been a systematic study of peptides in nervous and non-nervous tissues. The structure, distribution and some functions of in excess of 40 neuroactive peptides have been described, but it is not established whether all neuroactive peptides act as neurotransmitters. For some neuropeptides, many of the criteria for a substance being a neurotransmitter have been satisfied (Chapter 1). Details of synthesis, storage, transport, release and inactivation of neuroactive peptides can differ in detail from the analogous process for classical neurotransmitters, and these processes are discussed in general and illustrated schematically in Fig. 9.1.

Neuroactive peptides have been found to work in at least three different ways in the body. They appear to act as local neurotransmitters on

CLASSICAL NEURONE   'PEPTIDE' NEURONE

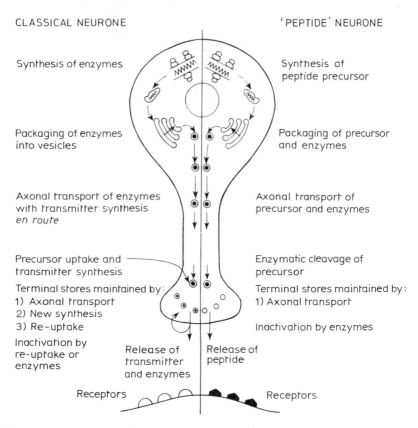

Synthesis of enzymes   Synthesis of
                                                     peptide precursor

Packaging of enzymes   Packaging of precursor
into vesicles   and enzymes

Axonal transport of enzymes   Axonal transport of
with transmitter synthesis   precursor and enzymes
en route

Precursor uptake and   Enzymatic cleavage of
transmitter synthesis   precursor

Terminal stores maintained by: Terminal stores maintained by:
1) Axonal transport   1) Axonal transport
2) New synthesis
3) Re-uptake   Inactivation by enzymes

Inactivation by   Release of   Release of
re-uptake or   transmitter   peptide
enzymes   and enzymes

Receptors   Receptors

Figure 9.1 Processing of neurotransmitter in a conventional (classical) neurone, and a peptide neurone (compare this figure with Fig. 1.1).

closely adjacent neuroeffector structures and they act also as local neurohormones, being released into the local portal systems as in the hypophyseal–pituitary system. Thirdly, they appear to act as hormones in their own right, being released into the bloodstream to act at sites distant from their point of release. Such diverse distribution methods and successively more diverse sites of action have suggested a large number of biological functions for these substances, including control of complex behavioural responses and pain pathways which require the concerted and integrated action of many different physiological and anatomical systems. Some of these functions will be discussed when the biology of the individual neuropeptides is described.

9.1.1 Classification of peptides

At least 40 neuroactive peptides have been identified in mammalian tissues, each having either a clearly defined or proposed neurobiological role; neuronal pathways have been described for only comparatively few neuropeptides and information about distribution is frequently incomplete.

There is no universally accepted classification system for the neuropeptides. One system which can be used is based on the major (or first described) anatomical location, and this sub-divides biologically-active peptides into four groups.

1. Neuropeptides found in the gastro-intestinal tract and CNS.
2. Neuropeptides of the hypothalamus which control release of anterior pituitary hormones.
3. Neuropeptides found in the anterior pituitary gland.
4. Neuropeptides found in the posterior pituitary gland.

Only those peptides which are most commonly located in the gut and brain and which have a proposed neurotransmitter role are discussed further in this chapter. Peptides associated with the hypothalamus and pituitary gland have a neurohormonal role and are dealt with in texts of physiology. A list of proposed neuropeptides and neurohormones is presented in Table 9.1, but this serves only as a guide and is by no means exhaustive.

9.2 SYNTHESIS, STORAGE, TRANSPORT, RELEASE AND INACTIVATION OF NEUROPEPTIDES

Peptides are synthesized in the nerve-cell body. The appropriate DNA is coded into mRNA, which is used as the template for ribosomal peptide synthesis. The peptide so produced is called the pre-propeptide, as it contains not only the propeptide (the amino acid sequence from which the neuropeptide is derived as explained below), but also a sequence of amino acid residues which act as signals for processing of propeptide into its storage form. The signal peptides are as yet little understood, but they appear to be important for ensuring that the pre-propeptide is delivered from the endoplasmic reticulum to the Golgi apparatus, where the pre-sequence is detached. The Golgi apparatus packages the propeptide into granules which are suitable for transportation to the nerve terminal in the microtubule system. The propeptide is stored in the terminal, but the process which controls its cleavage into peptide for release is not well understood. Enzymatic cleavage of the propeptide between pairs of basic

Table 9.1 Some proposed peptide neurotransmitters (and neurohormones) and their sites of origin

I	**Opioid peptides (located in CNS)**
	Leucine and methionine enkephalin
	β-Endorphin
	Dynorphin
II	**Gut and brain peptides**
	Cholecystokinin (CCK)
	Vasoactive intestinal polypeptide (VIP)
	Substance P (SP)
	Neurotensin (NT)
	Bombesin
	Gastrin
	Secretin
	Neuropeptide Y
	Neurokinin A and B
	Calcitonin gene-related peptide (CGRP)
	Angiotensin
III	**Hypothalamic peptides**
	Somatostatin
	Gonadotrophin releasing hormone (GnRH)
	Thyrotrophin releasing hormone (TRH)
	Bradykinin
	Growth hormone releasing factor (GHRF)
	Corticotrophin releasing factor (CRF)
IV	**Pituitary peptides**
	Anterior pituitary
	Corticotrophin (ACTH)
	Growth hormone (GH)
	β-Lipotropin
	Melanocyte-stimulating hormone (MSH)
	Thyroid-stimulating hormone (TSH)
	Posterior pituitary
	Oxytocin
	Vasopressin (anti-diuretic hormone) (ADH)

amino acids (arginine, lysine, serine) is common. These propeptide converting enzymes are as yet poorly characterized, but they could be important targets for drug action (e.g. captopril, section 3.4.4). In common with all neurotransmitters, the process of release of neuropeptides is calcium dependent. The relationship and time scale of release to breakdown of propeptide is not known.

It is generally believed that peptides act at specific receptors located on neuroeffector tissues, but to date only a small number of peptide receptors have been characterized. The task of pharmacological identification and characterization of peptide receptors depends on the availability of specific and selective agonists and antagonists, and to date few are available.

Termination of the activity of neuropeptides appears to be enzymatic; active transport uptake mechanisms have not been identified for peptides. The peptidase enzymes which inactivate peptides are not well characterized. The discovery of selective peptidase inhibitors may lead to drugs which will be able to prolong the actions of neuropeptides.

In several instances, fragments of neuropeptides have been found to have well-defined biological activity. Whether such peptides are important *in vivo*, and whether they are formed by the action of specific enzymes remains to be established.

9.3 CO-EXISTENCE AND CO-TRANSMISSION OF PEPTIDES WITH CLASSICAL NEUROTRANSMITTERS

Up until a decade ago, the prototypical neurone was believed to release one single classical neurotransmitter according to Sir Henry Dale's principle. Modern immunohistochemical techniques and radiolabelled ligand receptor binding studies now often demonstrate a marked overlap in the distribution of the classical neurotransmitters with the novel peptides. Indeed the presence of both types of neurotransmitter in the same nerve terminal has challenged Dale's hypothesis, suggesting that more than one transmitter can co-exist within neurones and they may indeed be released together in response to nerve stimulation and probably regulate the functional role of one another. The full functional significance of this phenomemon, called **co-transmission**, is not known. However it is suggested that one component acts as a true neurotransmitter by combining with the postsynaptic receptor site and initiating the postsynaptic response while the second component acts as a neuromodulator to alter the responsiveness of the transmitter–receptor interaction.

There are now a number of examples of classical neurotransmitters existing within peptidergic neurones (or vice versa); there are also examples of neurones that contain more than one proposed peptide neurotransmitter or more than one classical neurotransmitter. Examples of classical / peptide neurotransmitter co-transmission include dopamine and cholecystokinin within the ventral mesencephalon and mesolimbic forebrain terminals; 5-hydroxytryptamine and substance P in brainstem and spinal cord; and acetylcholine and vasoactive intestinal polypeptide (VIP) in salivary gland (section 9.7.1). Examples of two proposed neuropeptides in co-transmission are dynorphin and vasopressin; and

Table 9.2 Examples of neurotransmitters existing in co-transmission

Neurotransmitter 1	Neurotransmitter 2	Site
5-Hydroxytryptamine	Substance P CCK Enkephalin TRH	Brainstem; spinal cord
Acetylcholine	VIP Enkephalin	Cerebral cortex; salivary glands
	Substance P	Spinal cord Pons
Dopamine	CCK Neurotensin	Ventral mesencephalon
Noradrenaline	Enkephalin Neuropeptide Y	Brainstem cerebral cortex
GABA	CCK Somatostatin Substance P Enkephalin	Cortex; hippocampus, hypothalamus, basal ganglia
Adrenaline	Substance P Neuropeptide Y	Brainstem
GABA	5-HT Dopamine Glycine Glutamate Acetylcholine	Raphe nuclei Arcuate nucleus Cerebellum Cerebral cortex Medial septum

vasopressin and corticotrophin releasing factor: two classical neuro-transmitters identified together in the same neurone are GABA and 5-HT; GABA and glycine in the cerebellum. Refer to Table 9.2 for further examples.

9.4 OPIOID PEPTIDES

The opioid peptides are the most widely studied group of neuropeptides whose cellular distribution and functional roles focus on the CNS and gastro-intestinal tract. The discovery in 1975 by Hughes and Kosterlitz of two small peptide sequences, the enkephalins, which possessed potent morphine-like activity was the stimulus to a massive research effort. Since that time many other endogenous and synthetic compounds which interact at proposed opioid receptor sites have been discovered and the neurobiology of this complex system is slowly being documented. At present, there appear to be three distinct chemical families of naturally-occurring opioids – the enkephalins, the endorphins and the dynorphins.

Together these groups contain at least 10 endogenous opioid peptides which are biologically active, but distinct pharmacological or physiological roles cannot be ascribed to them all.

9.4.1 Synthesis

In keeping with the three major groups of endogenous opioids, three precursor molecules have been identified: pro-opiomelanocortin (POMC), pro-enkephalin and prodynorphin. These are all peptides with molecular weights of 25–30 000 and each contains a number of biologically-active peptides (both opioid and non-opioid (hormonal)). Thus POMC contains sequences for melanocyte-stimulating hormone (γ-MSH); adrenocorticotropin (ACTH); and β-lipotropin (β-LPH) – this latter structure containing the sequences of β-endorphin and β-MSH. Pro-enkephalin contains the structure for methionine-enkephalin (met-enkephalin); leucine-enkephalin (leu-enkephalin) and a residue octapeptide. Prodynorphin holds base sequences for leu-enkephalin, dynorphin A (1–17) which in turn can be further degraded to dynorphin A (1–8); dynorphin B (1–13); and α- and β-neo-endorphin fragments (Table 9.3). Within these precursor molecules the biologically-active peptide sequences are separated by pairs of basic amino acids

Table 9.3 Precursors of endogenous opioid peptides

Precursor molecule	Opioid and non-opioid peptides
Pro-opiomelanocortin (POMC)	γ-melanocyte stimulating hormone (γ-MSH) ACTH $\quad\rightarrow\alpha$-MSH β-lipotropin $\rightarrow\beta$-endorphin $\quad\quad\quad\quad\quad\alpha$-lipotrophin $\rightarrow\beta$-MSH $\quad\quad\quad\quad\quad\alpha$-MSH
Pro-enkephalin	Methionine-enkephalin Leucine-enkephalin Octapeptide
Pro-dynorphin	Leucine-enkephalin Dynorphin A (1–17) $\rightarrow$ Dynorphin A (1–8) Dynorphin B (1–13) α-neoendorphin β-neoendorphin

Base sequences
Methionine-enkephalin: Try-Gly-Gly-Phe-Met-COOH.
Leucine-enkephalin: Try-Gly-Gly-Phe-Leu-COOH.
Dynorphin A (1–8): Try-Gly-Gly-Phe-Leu-Arg-Arg-Ile-COOH.
α-Neo-endorphin: Try-Gly-Gly-Phe-Leu-Arg-Lys-Try-Pro-Lys-COOH.

which presumably act as markers for specific peptidases which cleave the opioid peptide from the precursor molecule.

9.4.2 Storage, transport, release and inactivation

The processing machinery from synthesis of the propeptide at the ribosomes through to packaging, transport, storage and cleavage of the biologically-active opioid peptide is not well understood. Similarly mechanisms of peptide release are only poorly documented, although the release of enkephalins in response to depolarizing stimuli applied to cerebral and spinal cord preparations is well reported, and this process appears to be calcium dependent. Drugs which can selectively affect these presynaptic processes are not available. Similarly the peptidases which inactivate synaptically-released or circulating opioid peptides have not been fully characterized; uptake processes for these molecules do not appear to exist.

9.4.3 Opioid receptors

The existence of multiple opioid receptors has been demonstrated by pharmacological bioassay techniques and by radiolabelled opioid binding studies. These experiments have clearly shown that not all opioid compounds exhibit the same expected agonist series potency, and the sensitivity to antagonists also varies between tissues. Such observations have led to the suggestion that there may be several types of opioid receptors with differing sensitivities for opioid agonists and antagonists. Current teaching thereby proposes multiple opioid receptors with possible further classification into various receptor sub-types. Within the CNS, three major opioid receptor subtypes are proposed (there also being a possible fourth), being assigned the Greek letters μ- (mu); δ; (delta); and κ- (kappa). (The other proposed receptor is σ- (sigma), but its agonist and antagonist drugs and functional roles are poorly defined.) These receptor sites vary not only in their affinity for specific agonists (although selective antagonists are not yet available), but also in their anatomical distribution and proposed function/pharmacological roles.

(a) μ-Opiate receptors

These receptor sites show a potency series in which β-endorphin is most potent, and morphine, methionine-enkephalin and leucine-enkephalin are successively less potent. μ-Opiate receptors are widely distributed in both the peripheral and central nervous systems, with particular high density in brainstem, and trigeminal nuclei, spinal cord, periaqueductal grey region, caudate-putamen, amygdala and cerebral cortex. In general, electrophysiological or pharmacological stimulation of μ-receptors induces hyperpolarization, by opening K^+ channels and closing Ca^{2+} channels, or by inhibition of presynaptic mechanisms as on primary afferent neurones in the dorsal horn of the spinal cord (Fig. 7.3). In keeping with this general suppressant action, some μ-receptors appear associated with the inhibition of the adenyl cyclase system resulting in reduced production of cyclic AMP. However at some sites μ-receptors are linked to excitatory actions, for example in hippocampus where μ-receptor mediated presynaptic inhibition on GABA-ergic nerve terminals results in inhibition of GABA release so evoking an overall excitatory response. In the peripheral nervous system activation of μ-opioid receptors in guinea-pig ileum results in inhibition of acetyl-choline release. The action of **morphine** to cause decreased pulse rate, slowing of peristalsis, constriction of pupils, respiratory depression, decreased response to painful stimuli and withdrawal syndrome after dependence appear to be μ-receptor mediated effects (Table 9.4). Thus withdrawal symptoms can be reversed by **morphine** itself or other μ-opioid receptor agonists but not by **ketocyclazocine** (or other κ-opioid receptor stimulants). β-Endorphin is proposed as the naturally-occurring ligand for μ-receptors. μ-Opioid receptors are more sensitive to the blocking actions of naloxone than are δ- or κ-opioid receptors.

(b) δ-Opiate receptors

These receptor sites show a potency series in which leucine-enkephalin is most potent, and methionine-enkephalin, β-endorphin and morphine are successively less potent. δ-Receptors are located in both the central and peripheral nervous systems; their distribution in general corresponds to tissue enkephalin concentrations. Overall δ-opioid receptor stimulation mediates hyperpolarization similar to that of μ-receptors. δ-Opiate receptors cause inhibition of release of noradrenaline from the nictitating membrane of the cat and from mouse vas deferens. δ-Receptors are less sensitive to the actions of naloxone, thus higher concentrations of this drug are required to block the actions of opioid agonists at this receptor site.

(c) κ-Opiate receptors

These receptor sites differ from μ- and δ-opiate receptors as κ-opiate receptor agonists cannot reverse the symptoms of **morphine** withdrawal in **morphine** dependent animals, and have a different spectrum of analgesic activity. κ-Opioid receptor agonists include pentazocine,

Table 9.4 Opiate receptors

	μ-Opiate receptor	κ-Opiate receptor	δ-Opiate receptor
Associated functions	Supra-spinal (cerebral) analgesia Respiratory depression Tolerance, withdrawal Physical dependence Euphoria; sedation	Spinal analgesia Hallucinations Tolerance Appetite suppression Sedation	Spinal analgesia Affective behaviour Tolerance
CNS distribution	Spinal cord Brainstem Periaqueductal grey Basal ganglia Cortex Amygdala	Spinal cord Brainstem Hippocampus Basal ganglia	General
Agonists	β-Endorphin (potent) **Morphine** **Met-enkephalin** Leu-enkephalin (weakest)	**Bremazocine** **Pentazocine** **Ethylketocyclazocine** **Butorphanol** Nalorphine	**Leu-enkephalin** (potent) **Met-enkephalin** **β-Endorphin** Morphine
Partial agonists	**Buprenorphine** **Meptazinol**		
Antagonists	Naloxone	Naloxone	Naloxone Ketocyclazocine
Selective agonist	**Sufentanyl**		
Selective antagonist	**Naloxonazine**	**Naltrindole**	**Met-enkephalin** Leu-enkephalin
Endogenous ligand	β-Endorphin	I–I3 fragment of dynorphin α, b-neoendorphin	
Associated ion channels, cyclase systems	K^+-ion channel adenylate cyclase system	Ca^{2+}-ion channel	Ca^{2+} and K^+-ion channels Inhibition of adenylate cyclase
Autonomic functions	Mydriasis/miosis and gut motility Respiratory depression Bradycardia Catecholamine inhibition	Respiratory depression Hypertension Catecholamine inhibition	Bradycardia

ethylketocyclazocine, bremazocine and butorphanol. These compounds cannot prevent pain induced by heat stimuli, while still diminishing perception of pressure- or chemically-mediated pain. They are believed to act predominantly at spinal sites, and thereby do not cause significant respiratory depression at doses which produce a similar degree of analgesia as other opioid agonists. They are however more liable to cause hallucinations and sedation than are other opioid agonist drugs. Electrophysiologically κ-receptor stimulation causes differing responses depending upon the anatomical site – excitatory actions being recorded in spinal cord, hippocampus and caudate-putamen; inhibitory effects at brainstem nuclei. **Dynorphins** (particularly the 1–13 fragment of dynorphin) and **neoendorphin** appear to be the endogenous agonists at κ-opioid receptors. **Naloxone** acts as a competitive antagonist at κ-opioid receptors but it is much less potent at these sites than at μ-opioid receptors.

9.4.4 The distribution and functions of opioid peptides

Once the opioid peptides had been characterized, it was possible to study their anatomical distribution both in the CNS and in the peripheral nervous system.

The distribution of opiate receptors is not uniform in the CNS, but it often complements the distribution of enkephalins. This association of high levels of potential neurotransmitter and its receptors is highly reminiscent of identified neurotransmitters.

Regions of the CNS with a rich enkephalin-like immunoreactivity and a high concentration of opiate receptors include the dorsal horn of the spinal cord, the peri-aqueductal grey matter and medial parts of the thalamus. These regions contain the pathways of the anterolateral (paleospinothalamic) tract, which conducts dull diffuse (as opposed to sharp localized) pain. It is suggested that opiate analgesics act on receptors in these regions to decrease pain.

High concentrations of enkephalins and opiate receptors are found in the amygdala and the frontal cortex, in parts of the hypothalamus and in the basal ganglia. These regions (with the exception of the basal ganglia) are associated with perception and expression of emotion, and form part of the limbic system. It is possible that decreased perception and response to pain is mediated in this region. The euphoriant effects of opiates may also be controlled from the limbic system; if so, then dependence might be controlled from this region.

The brain stem and medulla, the vagal nuclei and the area postrema have high concentrations of both enkephalins and opiate receptors. It is suggested that the antitussive (cough-suppressing) actions and nausea and vomiting effects of opiates are controlled in these regions. The enkephalinergic innervation of the region of the parabrachial nucleus may be involved in the respiratory depressant action of opiates.

Endorphins have been found in the pituitary gland and in the adrenal medulla. It is suggested that they may have a role in modulating pain during stress responses, at which time they are released both from the pituitary gland and from the adrenal medulla, along with ACTH from the former, and catecholamines from the latter. Endorphins have potent analgesic actions, and electrical stimulation of the peri-aqueductal grey matter results in an increase in the CSF concentration of β-endorphin. Patients suffering from chronic pain have been found to have abnormally low levels of β-endorphin in their CSF; values nearer to those found normally can be achieved in these patients by electrical stimulation, which in addition brings about analgesia.

Experimental evidence also suggests that endorphins may be involved in some forms of epileptiform seizures, while other experiments have led to suggestions that endorphin systems are important in reward. The idea of reward is important in psychology, and a chemically mediated reward system would fit in with observations on the development of withdrawal symptoms, and have implications for interpretation of the phenomenon of dependence.

Enkephalins and opiate receptors are found in the gastro-intestinal tract, a site where **morphine** reduces peristaltic propulsive movements of the small and large intestine. The constipating effect of morphine-like compounds leads to their use in the control of diarrhoea.

How opiates and opioid peptides bring about responses in neuro-effector tissues is not established; both neurotransmitter and neuro-modulator roles would be possible. In the gastro-intestinal tract, for example, **morphine** acts presynaptically to reduce acetylcholine release, and this probably contributes to decreased intestinal motility, but spasm of smooth muscle following **morphine** cannot be explained by a reduction of acetylcholine release; thus, other actions of **morphine** are indicated, possibly at postsynaptic sites. In the spinal cord, **morphine** and **enkephalins** have been shown to reduce the release of substance P from the dorsal horn. This appears to be a presynaptic action on specific opiate receptors found on substance P-containing neurones, which form part of the small-diameter primary afferent sensory neurones of the spinothalamic pathway (Fig. 9.2).

9.4.5 The pain pathways

Pain is interpreted as an unpleasant noxious stimulus, provoked by actual or potential tissue damage. Much research has been directed towards our

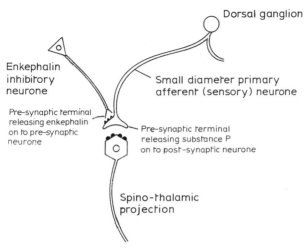

Dorsal ganglion

Enkephalin
inhibitory
neurone

Small diameter primary
afferent (sensory) neurone

Pre-synaptic terminal
releasing enkephalin
on to pre-synaptic
neurone

Pre-synaptic terminal
releasing substance P
on to post-synaptic neurone

Spino-thalamic
projection

Figure 9.2 One of the sites at which opiates may modify pain transmission (see also Fig. 9.3).

understanding of the physiological mechanisms involved in the pain response, the anatomical course of the pain pathways and the neurotransmitters associated with their function. The importance of this latter aspect has allowed the development of drugs to interact at pain synapses and thereby induce analgesia.

Pain signals generated from pain sensory endings (nociceptors) buried in the periphery and internal body organs travel along sparcely myelinated A δ or unmyelinated C primary afferent fibres of the dorsal root sensory neurones to the spinal cord; conduction of pain signals in C fibres is slow.

Within the spinal cord two ascending pain pathways are recognized with which the primary afferent sensory fibres synapse. One tract is the lateral spinothalamic pathway which conducts pain and temperature information from the spinal cord to the thalamus; the other is the spinoreticular system conducting pain and light touch sensation to the reticular formation of the brainstem and midbrain. Within the spinal cord the primary afferent sensory fibres of the dorsal root synapse in the outer layers of the dorsal horn (lamina II, the substantia gelatinosa). Interneurones may then cross over to the opposite side of the spinal cord to synapse with second-order spinothalamic neurones. These neurones then transmit noxious stimuli (pain and temperature) to the brain and terminate in the thalamus (ventral posterolateral nucleus): further third-order neurones project from thalamus to sensory cortex in the thalamo-cortical tracts. The spinoreticular tract projects pain and probably visceral sensory information from the posterior (dorsal) horn of the spinal cord to the reticular formation of the medulla and midbrain and then onto thalamus and sensory cortex through a multi-synaptic pathway.

It can be seen that the ascending pain pathways contain a number of synapses where sensory information can be modified. The neuro-transmitter candidate(s) of the primary afferent fibre for the transmission of noxious stimuli within the dorsal horn is presumed to be either an excitatory amino acid (glutamate or aspartate) or a neuropeptide (substance P), although other peptides such as somatostatin, VIP and CCK have also been proposed. However, within the dorsal horn a number of intrinsic interneurones have been identified, some of which employ the opioid peptides as neurotransmitters, in particular the enkephalins and dynorphins. Other neurotransmitters with a probable functional role at this site include GABA, glycine, 5-HT and other non-opioid peptides.

(a) The gate control theory

In addition to the two major ascending spinal pain conducting pathways described above, a number of descending systems are believed to exist which help to control or modify pain transmission. Midbrain and brainstem sites contain opiate receptors which when stimulated appear to have analgesic properties. It is known that neurones project from sites such as raphe nuclei, periaquaductal grey matter and reticular formation down the spinal cord (in corticospinal and reticulospinal tracts) to terminate on dorsal horn interneurones. These descending pathways may themselves release enkephalins, and 5-HT (from raphe nuclei) or noradrenaline (from locus coeruleus).

The general effects of both the interneurones within the dorsal horn (substantia gelatinosa) and of the descending regulatory pathways is believed to filter or to gate control the incoming nociceptive signals at this level in the spinal cord. Thus it is suggested that opioid peptides released at this site serve to modulate incoming nociceptive information by their action on μ-opiate receptors located on the terminals of the primary afferent fibres. Indeed **enkephalin** and **morphine** have elegantly been shown to decrease substance P release in response to stimulation of dorsal root fibres in experimental preparations (Fig. 9.2) Secondly, sectioning of the dorsal root fibres (rhizotomy) greatly reduces the number of μ-receptors and enkephalin binding sites, as well as reducing substance P content of the dorsal horn of the spinal cord. 5-HT release at this site has been postulated to reduce incoming innocuous signals with the overall effect of enhancing the sensation of noxious stimuli. This mechanism of modulating nociceptive stimuli by neuronal and neurotransmitter interactions is known as 'the gate control theory of pain regulation'; the action of some fibres is to close the gates to pain, while others function to open the gates to pain.

(b) Pain and analgesia

Clinically, pain can be divided into two categories – acute pain associated with acute cellular injury (e.g. trauma, myocardial infarction), and chronic pain (e.g. progressive organ damage, malignancy). In the management of any disease state, the control of pain is of paramount importance to the clinician even if the underlying pathology cannot be reversed. The physician is equipped with an armoury of pain-killing drugs which fall into four general groups:

1. Opioid-like drugs;
2. Non-steroidal anti-inflammatory drugs (NSAIDs);
3. Local anaesthetics;
4. General anaesthetics.

There are also a number of miscellaneous agents which can be employed in the control of neuralgic pain but for which the mechanism of action or site of interaction along the anatomical pain pathways are not fully elucidated; for example **carbamazepine** can be successful in relieving neuralgic pain and this drug may function at the level of the thalamus (and therefore be effective in thalamic syndrome). **Amitriptyline**, a tricyclic antidepressant agent, in low doses has been reported as having analgesic activity: its mode of action in this role is unclear but suggestions have included enhancing noradrenergic transmission within the gate control mechanism in the spinal cord. Benzodiazepines are occasionally used when anxiety coexists with pain.

A new clinical approach for the control of chronic pain currently undergoing appraisal is the use of infusions of high doses of **naloxone** which appears to have a long-lasting analgesic effect: the mechanism of this action requires further investigation. Another procedure which has a use in pain control is transcutaneous nerve stimulation whereby low-intensity nerve stimulation produces a local analgesic effect which is proposed as being related to release of endogenous opioids. Similarly it has been suggested that the oriental practice of acupuncture to induce analgesia might depend for its effectiveness on the stimulation of pain-modulating enkephalinergic pathways. This idea is supported by the observations of increased levels of enkephalins in the cerebrospinal fluid (CSF) following acupuncture, or following electrical stimulation of the periaqueductal grey matter in the brainstem. **Naloxone** can block the analgesia induced either by acupuncture or electrical stimulation.

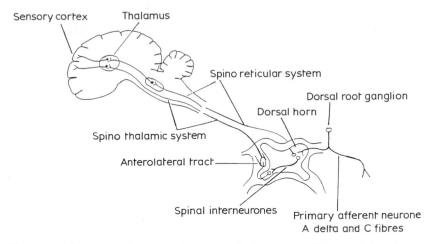

Figure 9.3 The pain pathways and sites at which pain could be modulated.

A study of the anatomical pathways involved in pain transmission (Fig. 9.3) will indicate that effective analgesics may act at differing levels to control the sensation of pain, for example at the sensory nerve terminal, within the spinal cord, at the level of the brainstem or thalamus, and even at the sensory cortex. Thus tissue injury as a cause of pain with the local release of a number of biologically-active compounds such as the kinins (bradykinin, kallidin, histamine and prostaglandins) which evoke an inflammatory reaction, may respond well to NSAIDs (e.g. propionic acid derivatives – **ibuprofen, naproxen, fenbufen**; the salicylates – **aspirin**; and other agents such as **paracetamol, indomethacine, mefanamic acid, phenylbutazone**): this situation occurs in arthritis and other musculo-skeletal disorders.

Local anaesthetics are administered topically or by injection for (as their name implies) local pain relieving effects. They work by blocking ion channels in nerves (as well as on other electrically excitable membranes). In nerves they block cationic channels (channels for Na^+, K^+, Ca^{2+}) and thus reduce axonal conduction (section 1.3). If axonal conduction is blocked in sensory neurones, then local analgesia will result; if axonal conduction is blocked in motor nerves, then motor paralysis will occur; block of axonal conduction at axon terminals will inhibit neurotransmitter release.

General anaesthetics are given systemically either by inhalation or intravenous injection, and they produce surgical anaesthesia by actions in the CNS. Their mechanism(s) of action are not well understood; theories of action include interactions with lipid, protein or water components of neuronal membranes (leading to stabilization of the membrane) rendering it unresponsive to physiological stimulation.

Moderate to severe pain, particularly of visceral origin, may require stronger analgesics in the form of opioid drugs (narcotic analgesics). Drugs of choice include **morphine, diamorphine, pethidine, codeine, buprenorphine, dextropropoxyphene** and a number of others. These are considered in section 9.4.6 below. The use of slow-release opioid preparations in the management of terminal care is now well established and such drugs should not be denied patients with chronic incurable pain for fear of dependence or addiction.

9.4.6 Therapeutic applications of drugs acting at opiate receptors

Drugs which interact with opiate receptors can be considered to act either as agonists or antagonists, depending on the therapeutic use to which they are put. **Morphine, heroin, pethidine, methadone, codeine, levorphanol, pentazocine** and **buprenorphin** are used as opiate agonists. **Nalorphine, levallorphan** and **naloxone** are used as opiate-receptor blockers (antagonists). Several compounds in the above lists, however, are partial agonists; thus, **pentazocine, buprenorphin, nalorphine** and **levallorphan** have mixed agonist and antagonist properties. The compounds listed above are used for a wide range of actions, and their uses are dependent on considerations such as the efficacy, potency and duration of action. Dependence is a problem which can occur with all opiates that have agonist activity. It is hoped that compounds which retain desirable properties such as analgesia, but are free of dependence-producing actions, might one day be developed.

Morphine is used as an analgesic which has tranquillizing actions. It can induce a relaxed sleep, and reduce fearfulness associated with pain. It can be given orally or parenterally, and is of most value in relieving dull prolonged pain, while being relatively ineffective in controlling sharp pain. **Morphine** reduces coughing and can decrease diarrhoea. Other actions of **morphine** can be regarded as undesirable, and are part of the price for obtaining pain relief and sleep.

(a) Side effects

These increase with increasing plasma concentrations of **morphine**. Respiratory depression and a decreased response to the build-up of carbon dioxide in the blood can lead to death as a result of respiratory arrest. Nausea and vomiting occur in a significant proportion of patients and, while this can be controlled with anti-emetics (phenothiazines or anticholinergics), the suppression of the cough reflex may allow inhalation of vomit and contribute to the decreased respiratory function. Spinal reflexes are increased by **morphine**, and convulsions can occur. The pupils constrict following high doses of **morphine. Morphine** has weak direct bronchoconstrictor activity, but life-threatening broncho-constriction can occur in asthmatic patients if **morphine** is given intravenously, when it can cause histamine release. Constipation occurs with prolonged use of **morphine**, and spasm of smooth muscle in the biliary tract can lead to colic pains.

(b) Tolerance

Most of the actions of **morphine**, except those on the gastro-intestinal tract and the eye are subject to the development of tolerance. It remains to establish the mechanism(s) by which tolerance occurs but, as with all examples of tolerance, higher doses have to be given to obtain similar responses. Cross-tolerance occurs to the action of other opiate agonists.

(c) Dependence

Because of dependence and the consequent physical and psychological withdrawal symptoms, the use of **morphine** and related drugs must be carefully controlled. Physical withdrawal symptoms in the dependent subject occur when the drug, or a substitute, is not available. They include anxiety, agitation, sleeplessness, vomiting and diarrhoea, abdominal and leg cramps, sweating, streaming of the eyes and nose, twitching of the muscles and possibly convulsions and also waves of pilo-erection (goose flesh). In severe dependence, withdrawal symptoms can last for 6–10 days. The severity of withdrawal symptoms is dependent on the degree of dependence, which itself varies according to the agent which caused dependence. Psychological dependence manifests itself as a desire for the drug, and this may persist for much longer than do the physical withdrawal symptoms.

(d) Overdose

Overdose of opiate agonists causes respiratory depression, is accompanied by pin-point pupils and leads to coma and death. If the symptoms of opiate overdose are recognized in time, it is usually possible to prevent death due to respiratory arrest by treatment with an opiate-receptor blocker such as **naloxone. Naloxone** is virtually free of opiate-agonist actions, and it will displace any agonist molecules from opiate receptors. It is given by intravenous injection, and respiratory function improves rapidly if the dose has been adequate. In opiate-dependent subjects, **naloxone** will precipitate withdrawal symptoms. Respiratory depression due to agents other than opiates will not be reversed by **naloxone. Naloxone** has a duration of action shorter than that of many opiate agonists, and thus repeated administrations may be needed to prevent the return of respiratory depression due to the presence in the plasma of opiate agonist. The analgesic and other therapeutically desirable actions of opiates are also reversed by **naloxone**.

Whereas **naloxone** is an opiate-receptor blocker, **levallorphan** and **nalorphine** (which are used as opiate antagonists) have some actions which mimic those of opiate agonists. Thus, whereas they are used in the treatment of overdose with opiate agonists, and can successfully antagonize opiate-induced respiratory depression, they can, owing to their partial agonist actions at opiate receptors, exacerbate respiratory depression due to ethanol or barbiturates. For this reason, **naloxone** is a generally safer alternative.

Other opiates differ from **morphine** by virtue of their varying potency and efficacy (and therefore dependence-producing potential), and some degree of selectivity in the site of action.

Heroin (diamorphine) is more lipid soluble than **morphine** and, following intravenous injection, higher concentrations occur in the CNS. It is favoured among addicts as it produces a better 'high'. Dependence occurs more readily with **heroin** than with other opiates. It is generally used only for pain relief in terminal illness.

Methadone is of approximately the same potency as **morphine**, but has a longer duration of action. It causes less euphoria than does **morphine**, and the dependence potential and withdrawal symptoms are less than with **morphine**. It is used as a substitute for **morphine** or **heroin** during withdrawal of these drugs from dependent subjects.

Pethidine is used when pain is not severe enough to warrant the use of **morphine**, but is more severe than that which can be controlled by **codeine. Pethidine** has weak hypnotic action, does not effectively suppress cough and is less likely to cause constipation. It is used in obstetrics and postoperatively, because it has a short duration of action. If high doses are given, **pethidine** can cause respiratory depression and, since it can cross the placental barrier, it can depress foetal respiration. As with all opiate agonists, dependence can occur. **Pethidine** has anti-muscarinic actions and thereby may cause dry mouth and blurred vision.

Fentamyl is similar to **pethidine** but shorter acting. It is used in anaesthesia to induce neuroleptoanalgesia.

Codeine (3-methylmorphine) is partially demethylated in the body to **morphine**, and is less potent at relieving pain than is **pethidine**. It has only weak respiratory depressant activity, and thus quite high doses can be given to suppress coughing and control diarrhoea.

9.5 SUBSTANCE P (THE TACHYKININS)

Substance P (SP) was the first neuropeptide to be described and studied, being originally extracted crudely as a white powder from equine gut and brain by von Euler and Gaddum in 1931. The results of these earlier studies must be interpreted with caution as the extracts would undoubtedly have contained many biologically-active substances in addition to SP. SP is an eleven amino acid residue peptide with the base sequence NH_2Arg-Pro-Lys-Pro-Gln-Gln-Phe-Phe-Gly-Leu-Met-COOH. SP is now considered as a member of a major class of neuropeptides exhibiting a wide range of potent biological actions called the tachykinins.

9.5.1 Synthesis, storage, release and metabolism

Details concerning the manufacture of SP are not fully available although it is suspected it is derived from larger precursor molecules α- and β-pre-protachykinin both of which contain the SP sequence. The packaging into vesicles, axonal transportation and storage mechanisms of SP are likewise only patchily understood. The potassium-stimulated release of SP has been demonstrated in a number of perfused tissues or cell preparations including spinal cord, dorsal root ganglia cells and substantia nigra. Release is calcium dependent. SP release from dorsal horn fibres can be reduced by the presence of **morphine**, an action sensitive to the blocking actions of naloxone (section 9.4).

Several enzyme systems, the endopeptidases, have now been described which efficiently degrade and thereby inactivate SP.

9.5.2 Substance P receptors

Current research proposes at least two sub-types of SP receptor, being identified by the activity of two proposed agonists – physalaemin (P) and eledoisin (E): hence SP-P and SP-E receptors. Radioligand binding studies have mapped out high densities of binding sites in different regions of the CNS including caudate-putamen, septum, amygdala, hippocampus, olfactory bulbs, pons, nucleus tractus solitarius and laminae I, II and X of the spinal cord including the substantia gelatinosa. There is little information currently available on the differential distribution of SP-P and SP-E sites. Selective SP receptor blocking drugs do not yet appear to be available although this is a field of active research.

9.5.3 Distribution of SP-like immunoreactivity in the nervous system

Immunohistochemical techniques indicate that SP-like substance is concentrated in areas of the nervous system which often show high population of binding sites, viz. substantia nigra, the nuclei of the trigeminal tracts, and in layers I, II and III of the Rexed system of the dorsal spinal cord. Lower concentrations of SP-like immunoreactivity are found in the hypothalamus, ventral tegmentum, amydala, nucleus accumbens, globus pallidus, olfactory tubercle, habenula and caudate-putamen: lowest concentrations are recorded in the frontal cortex, hippocampus and cerebellum.

Immunohistochemical mapping studies have identified some SP-like immunoreactivity containing pathways in the CNS. Within the basal ganglia a pathway projecting from the caudate-putamen to the substantia nigra pars reticulata has been demonstrated. In the spinal cord SP appears in association with sensory pathways of the dorsal root ganglia and substantia gelatinosa.

Outside the CNS, SP-like immunoreactivity is found in the terminals of sensory neurones, in the Auerbach's and Meissner's plexuses of the gut, in the walls of blood vessels and bronchi, and in the salivary glands where it induces secretion.

9.5.4 Functional roles of Substance P

The functional roles of SP have been suggested on the basis of the effects of administration of this compound to animals and on its anatomical distribution. Its major functional role appears in the regulation of sensory mechanisms.

SP causes pain if applied to the cut ends of sensory neurones. It has been suggested that it may be responsible for neurogenic pain and inflammation, and be the agent which brings about the late vasodilation phase of the Lewis Triple syndrome (section 6.4.1). Certainly the intradermal injection of SP produces a local flare and slowly developing wheal. The former reaction is probably related to release of histamine from mast cells, the latter response (wheal) may be due to SP itself. Within the spinal cord it is suggested that SP acts as a primary sensory neurotransmitter, being associated with known pain pathways. Certainly its release at this site appears to be modified by opioid agonists, for there are opiate receptors located on the SP terminals in the dorsal horn of the spinal cord (section 9.4). Within the basal ganglia SP fibres appear in close synaptic connection with dopamine cell bodies of the substantia pars compacta. Iontophoretic application of SP to these nigral cell bodies causes an increased rate of firing of the projecting dopamine fibres with subsequent activation of forebrain dopamine systems and related behavioural stimulation. SP may therefore have a role, by way of a descending strio-nigral feedback pathway, in the regulation of the ascending nigro-striatal dopamine system and the control of motor behaviour. In support of this concept, post-mortem studies have confirmed changes in concentrations of this neuropeptide in brains of patients suffering extrapyramidal movement disease (e.g. Parkinson's disease, Huntington's chorea).

If SP is injected systemically, it causes a fall in blood pressure due to the dominance of vasodilator actions. In the gut SP causes contraction of smooth muscle, and is believed to have a controlling function in peristalsis. In the respiratory tract, SP released from nerve fibres may cause bronchoconstriction and increased mucus secretion.

9.6 CHOLECYSTOKININ (CCK)

Whilst a 33 amino acid residue peptide (CCK 33) has a well defined role stimulating secretion of pancreatic enzymes and contraction of the gall bladder in the gastro-intestinal tract, a C-terminal octapeptide residue of cholecystokinin (CCK 8) has been identified in the brain and spinal cord. CCK 8 demonstrates a heterogeneous distribution within the mammalian CNS, with very high concentrations in cerebral cortex, hippocampus, the amygdala and septum. Significant amounts are found in periaqueductal grey matter, the dorsomedial nucleus of the hypothalamus, dorsal raphe nucleus, the caudate-putamen and ventral tegmentum. Binding studies identifying CCK receptors show that in general receptor density parallels endogenous CCK concentrations.

CCK-related peptides can be released in response to a depolarizing stimulus from synaptosomes or slice preparations of cerebral cortex, hypothalamus and periaqueductal grey in a calcium-dependent manner. In general the iontophoretic application of CCK 8 to cerebral neurones causes an excitatory action.

9.6.1 CCK receptors

Studies on the classification of CCK receptors have led to the development of several proposed CCK receptor antagonists (e.g. asperlicin, proglumide, benzotript), but these in general appear to have far greater potency within the gastro-intestinal tract than in the CNS. This difference of affinity has led to the proposal of two CCK receptor sub-types: pancreatic Type A and CNS Type B.

9.6.2 Functional role of CCK: co-transmission

Of particular interest in studies with CCK is the fact that this neuro-peptide may occur in co-existence with classical neurotransmitters in the brain. CCK distribution overlaps with the A_8–A_{10} dopamine-containing cell bodies of the substantia nigra, pars compacta and ventral tegmentum, and with the B_6 and B_8 5-hydroxytryptamine-containing cell bodies of the raphe nuclei. Similarly high concentrations of CCK-like immuno-reactivity occur in the projection areas of these cell groups, viz, caudate-putamen, amygdala, lateral septum and cerebral cortex. This anatomical distribution has led to speculation about a close functional interaction between CCK and dopamine (and 5-HT). Indeed experimental work has shown that this neuropeptide can either potentiate or inhibit actions of dopamine in the mesolimbic pathway (e.g. CCK can decrease dopamine release). Suggestions that CCK may function by modulating dopamine- (or 5-HT)-mediated neurotransmission appear valid. Other work has hinted at the involvement of CCK in nociception and feeding behaviour in animals although any functional relationships are still highly speculative.

9.7 VASOACTIVE INTESTINAL POLYPEPTIDE

Vasoactive intestinal polypeptide (VIP) is a 28 amino acid residue peptide, found in nerves in the alimentary canal from the oesophagus to the rectum, and in ancillary organs such as pancreas and gall bladder. It is also found in some blood vessels and in the bronchi. VIP relaxes most smooth muscle; thus it brings about vasodilation leading to hypotension, bronchodilation, and relaxation of intestinal muscle, which is not mediated by β-adrenoceptor mechanisms. In the pancreas, VIP causes release of insulin, glucagon and somatostatin, and in the adrenal cortex it stimulates steroidogenesis.

VIP-like immunoreactivity is found in many areas of the CNS including cerebral cortex, hypothalamus, amygdala, hippocampus, corpus striatum, periaqueductal grey and in some primary afferent neurones projecting to the spinal cord. VIP-like immunoreactivity is released in response to a depolarizing stimulus (potassium) from perfused hypothalamus slices, and the released product is rapidly degraded by peptidases. Detailed information on this peptide's synthesis and storage mechanisms is not yet available. When applied iontophoretically to CNS neurones, VIP usually causes depolarization and excitation. Radio-labelled ligand binding studies have identified two VIP binding sites – a high-affinity site and a low-affinity site. However, the significance of these presumed receptor sites remains to be established. It is believed that many of the pharmacological effects of VIP are mediated through an adenylyl cyclase system. VIP can stimulate the release of prolactin, ACTH, luteinizing hormone and growth hormone from the anterior pituitary; it has an inhibitory effect on the release of somatostatin. Selective agonists and antagonists for VIP receptor sites are not yet known.

The possible functional roles for VIP within the CNS have not been fully established. Its action on anterior pituitary hormone release may implicate a neuroendocrine role. As indicated above, VIP has a predominantly potent relaxant effect on smooth muscle, subsequently causing vasodilation and bronchodilation but the physiological relevance is to be established.

9.7.1 Co-transmission

In the cat salivary gland (and in some other tissues) VIP and acetylcholine (ACh) have been identified within the same neurones, and electrical stimulation of the post-ganglionic nerve fibres releases both ACh and VIP-like immunoreactivity. The effect of such stimulation is to cause both vasodilation and increased salivary flow. This latter effect appears to be cholinergically-mediated through muscarinic receptors as it can be blocked by **atropine**, but this drug has no action on the vasodilation which is presumably VIP-mediated. When VIP is present at the same time that ACh is stimulating secretion, however, the saliva flow is potentiated and prolonged. The mechanism for this potentiation of response to ACh could be the observation made during ligand binding studies that in the presence of VIP, the binding of muscarinic agonists is increased by several orders of magnitude. It has been suggested that there may be an allosteric VIP binding site on the muscarinic receptor, which increases the ability of muscarinic agonists (including of course ACh), to bind to the muscarinic receptor.

9.8 NEUROTENSIN

Neurotensin (NT) is a 13 amino acid residue peptide which was originally identified as a potent vasodilator. NT-like immunoreactivity has been demonstrated in many regions of the CNS, where it is often concentrated in axon terminals. Areas of high concentration in human brain include the hypothalamus, basal ganglia, the interstitial nucleus of the stria terminalis, the limbic system and the dorsal part (substantia gelatinosa) of the spinal cord. Neurotensin-like immunoreactivity is also located in the mucosal endocrine cells of the small bowel.

Neurotensin is probably formed from a large precursor molecule preproneurotensin. The axonal transportation and storage mechanisms of NT are not established. The potassium-evoked calcium-dependent release of NT-like immunoreactivity has been demonstrated from slices of perfused hypothalamus. Released NT is rapidly degraded to inactive products by peptidases. Electrophysiologically NT has been shown to have both excitatory and inhibitory actions on central neurones, depending on the anatomical site. In the gut this peptide causes depolarization of myenteric neurones.

NT binding sites have been identified using radiolabelled peptide, with high accumulation in substantia nigra, ventral tegmentum, thalamus and hypothalamus, median raphe nucleus and stria terminalis, and substantia gelatinosa of the spinal cord.

9.8.1 Functional role of neurotensin

The neurobiology of NT is not well understood, but like other neuropeptides, it has been associated with a number of possible functions within the CNS and gastro-intestinal tract. It has been implicated in the control of pain responses but its antinociceptive action is not blocked by the opioid antagonist **naloxone**. Anatomically the site of such action is not clear although medial thalamus, amygdala, periaqueductal grey and the substantia gelatinosa of the spinal cord appear able to mediate this effect. Like CCK, NT is proposed as having a close association with dopamine neurones (although not co-existence) and the peptide can modify the effects of dopamine agonists on motor activity: a modulatory role of NT on dopamine function within mesolimbic neurones is thereby suggested.

9.9 SOMATOSTATIN

Somatostatin (ST) is a 14 amino acid residue peptide although it is also recognized as a 28 unit residue. These have a widespread distribution throughout the CNS with highest concentrations in hypothalamus, hippocampus, cerebral cortex and limbic areas. Mapping studies have demonstrated ST-like immunoreactivity in fibre projections from the hypothalamus ascending rostrally to the limbic forebrain and projecting caudally to the substantia nigra and locus coeruleus. The spinal cord also contains high concentrations of ST-like immunoreactivity throughout its length, particularly lamina II of the dorsal horn. Some studies have suggested the co-existence of ST with other classical neurotransmitters, in particular noradrenaline and acetylcholine.

The calcium-dependent, potassium-evoked release of somatostatin 14 and 28 peptide has been demonstrated from hypothalamic slices. Radioligand binding studies report high concentrations of receptor sites within hippocampus and limbic regions of amygdala, stria terminalis, olfactory tubercle and septum; low concentrations of binding sites occur in cerebral cortex. In general electrophysiologically ST has an inhibitory action on most CNS cells evoking hyperpolarization.

9.9.1 Functional role of somatostatin

ST has a proposed functional role in sensory perception, being associated with the primary afferent sensory neurones of the spinal cord where it has been shown to inhibit substance P released in response to noxious stimuli. The precise mechanism of this proposed antinociceptive effect is to be established.

Physiological release of ST from the median eminence promotes the secretion of growth hormone from the anterior pituitary gland, although paradoxically this peptide reduces growth hormone secretion from pituitary adenoma cells. Thus a possible role of ST analogues in the management of acromegaly is proposed.

Its anatomical distribution in limbic forebrain has linked ST with control of mood and motor behaviour although a firm basis for this hypothesis is yet to be established.

9.10 BOMBESIN

Bombesin is a 14 amino acid peptide originally isolated from amphibian skin but more recently identified in mammalian CNS and gut. Bombesin-like immunoreactivity is demonstrable in the myenteric plexus of the intestine, in the substantia gelatinosa and nucleus tractus solitarius of spinal cord and brain stem, and in the mesolimbic regions of amygdala and nucleus accumbens. In general this peptide has a stimulant effect on smooth muscle (e.g. gut) and an excitatory action on cells in the CNS. Although the precise physiological and pharmacological roles of bombesin are not known, its anatomical locations and experimental observations have linked this peptide with possible functions of satiety, neurotransmission of sensory information within the spinal cord, stereotyped behaviour and motor activity.

FURTHER READING

Akil, H., Watson, S.J., Young, E., Lewis, M.E., Khachaturian, H. and Walker, M.J. (1984) Endogenous opioids: Biology and function. *Ann. Rev. Neurosci.*, 7, 223–55.

Bloom, F.E. (1983) The endorphin. *Ann. Rev. Pharmacol. Toxicol.*, 23, 151–70.

Campbell, G. (1987) Co-transmission. *Ann. Rev. Pharmacol. Toxicol.*, 27, 51–70.

Haynes, L. (1988) Opioid receptors and signal transduction. *TIPS*, 9, No. 9.

Hughes, J. (1975) Isolation of an endogenous compound from the brain with pharmacological properties similar to morphine. *Brain Res.*, 88, 295–308.

Iversen, L.L. (1983) Nonopioid neuropeptides in mammalin CNS. *Ann. Rev. Pharmacol. Toxicol.*, 23, 1–27.

Lewis, D.A. and Bloom, F.E. (1987) Clinical perspectives on neuropeptide. *Ann. Rev. Med*, 38, 143–8.

Low, L-M, and Pfaff., D.W. (1988) Neuromodulatory actions of peptides. *Ann. Rev. Pharmacol. Toxicol.*, 28, 163–88.

Maggie, J.E. (1988) Tachykinins. *Ann. Rev. Neurosci.*, 11, 13–28.

Smith, A.P. and Lee, N.M. (1988) Pharmacology of dynorphin. *Ann. Rev. Pharmacol. Toxicol.*, 28, 123–40.

Index